GREENLAND

EUROPE

ARCTIC OCEAN

NORTH ATLANTIC OCEAN

AFRICA

SOUTH AMERICA

INDIAN OCEAN

SOUTH ATLANTIC OCEAN

EXPLORING SPACE WITH A CAMERA

Nearly one full hemisphere of the Earth is revealed in this picture, showing South America, much of North America, and parts of Africa, Europe, and Greenland. It was taken on November 18, 1967, by ATS–III, which was stationed over the Equator at approximately 47° W. "This picture," explains LEONARD JAFFE, Director, Space Applications Programs, NASA, "is actually made up of 2400 horizontal lines, with each line encompassing a swath of Earth approximately 2 nautical miles wide and extending from horizon to horizon. Using very careful scrutinizing techniques, features as small as 2 nautical miles on a side can be seen. As evidenced here, color makes it easier to distinguish between clouds, dry land, green vegetation, and bodies of water. A particular example is the clarity with which Lake Titicaca (a tiny dark spot) can be observed at the bend in the west coast of South America."

NASA SP–168

EXPLORING SPACE
WITH A CAMERA

COMPILED AND EDITED BY EDGAR M. CORTRIGHT

Scientific and Technical Information Division
OFFICE OF TECHNOLOGY UTILIZATION   1968
NATIONAL AERONAUTICS AND SPACE ADMINISTRATION
Washington, D.C.

21185

FOR SALE BY THE SUPERINTENDENT OF DOCUMENTS, U.S. GOVERNMENT PRINTING OFFICE, WASHINGTON, D.C., 20402—PRICE $4.25

LIBRARY OF CONGRESS CATALOG CARD NO. 68-60027

FOREWORD

Down through history in no decade has man acquired such far-ranging new concepts of motion and its role in the universe, and so much new knowledge about his own planet, its natural satellite, the Moon, and its sister planets, Mars and Venus, as in the ten years just passed. This new knowledge feeds new processes of thought and new understanding of many scientific disciplines and broad categories of technology.

From positions in space and from machines in rapid motion, we have measured the Earth's weather phenomena and atmospheric dynamics; we have fixed the precise locations of mountains in Antarctica and lakes in Peru. We have mapped the Moon in detail and have made our first analyses of its composition. We have discovered and measured the Earth's trapped radiation belts and explored many facets of the solar wind. We have learned to use rockets to send instruments to measure the temperatures of Venus and to photograph the surface of Mars.

These ten years have often been likened to that period of the 15th century when man learned to traverse the oceans and use them to explore our planet Earth. But the pace of space exploration is faster and the implications for man's intellectual and physical development are more far-reaching.

In 1965 the Committee on Science and Astronautics of the U. S. House of Representatives published a report made by the National Aeronautics and Space Administration to President Johnson which summarized future opportunities opening up in aeronautics and in space. The following is quoted from that report:

> Down through the course of history, the mastery of a new environment, or of a major new technology, or of the combination of the two as we now see in space, has had profound effects on the future of nations; on their relative strength and security; on the relations with one another; on their internal economic, social, and political affairs; and on the concepts of reality held by their people.

The photographs in this book are a selection from the thousands returned to Earth during the decade. They were initially assembled by Edgar M. Cortright, a senior official of the National Aeronautics and Space Administration, as a personal reminder of the stirring years when man first developed the ability to propel his cameras and instruments, and then himself, beyond the Earth's atmosphere. With this publication, they become a part of NASA's flow of reports to the American public on some of the growing returns from its investment in the exploration of space. As such, I hope this book can find its way into many American homes, for it is a part of a record of achievement of which our country can be proud.

Photographs from space tell only part of man's reach through the air and on into space. Cameras cannot show the data gathered by many kinds of scientific sensors. They also fail to reflect—except inferentially—the immense impact that our air and space efforts have had on broad areas of life on Earth. And yet, through joining our efforts in aeronautics and space, we have made great strides in the advance of many technological areas. In 1958 we could scarcely put 100 pounds into Earth orbit; now the Saturn V can orbit 140 tons. At the beginning of the decade our lunar probes were failing; now we have reconnoitered the Moon with automated spacecraft and man is almost ready to follow. Ten years ago the utility of space was a gleam in the eyes of enthusiasts; now we have global weather analysis, communications satellites that link continents together, and geodetic and navigational satellites at the service of the whole world.

To do all this, public officials, legislators, scientists, engineers, technicians, and managers have had to tackle and solve new, complex, and demanding problems. Our nation's reach into the space environment is the most complex of the large non-military undertakings that man has yet attempted. Our country's policy is to do it for peaceful purposes for the benefit of all mankind. As this pictorial report portrays, all mankind has new material to use in moving rapidly to develop new concepts of the reality of the Earth, of the solar system, and of the dynamics and phenomena which control so much of our lives here on Earth. These remarkable photographs cannot reflect all, but they certainly show much of the results we have achieved.

JAMES E. WEBB, *Administrator*
National Aeronautics and Space Administration

CONTENTS

INTRODUCTION

THIS IS A COLLECTION OF the best photographs taken from space during the first decade of space exploration. The reader may miss the familiar pictures of launch pads, rockets, and tracking antennas. Here he will find instead views seen through the eyes of cameras in the eerie solitude of space—orbiting hundreds or thousands of miles above the Earth's surface, swooping low over heretofore unseen mountains and craters of the Moon, resting gently on the lunar plains, and gliding silently past the red planet Mars. Sometimes the cameras were pointed by men. Sometimes they were pointed by machines with computer brains, gyroscopic senses, star-seeking eyes, and servomechanism muscles powered by solar energy. The men brought their films back with them. The machines radioed their pictures back to an Earth to which they would never return. In both cases, they recorded views never before seen by men.

These historic pictures depict, far better than words, man's determined assault on the secrets of space. Some of the early pictures seem almost amateurish, taken by untried and relatively primitive equipment. But at the time they seemed wondrously exciting, and men of the space program would cluster about a freshly received picture and say to each other almost incredulously, "Look! There's England and it's just like the maps!"

As the years counted down, one problem after another was encountered and solved, and one objective after another achieved. "The difficult we do immediately, the impossible takes a little longer" seemed to fit the space program. Reading about a new success began to seem routine to some. But new pictures of new sights still drew excited clusters of men and the exclamations:

"Look! There's the whole Earth!"

"Look! The back of the Moon doesn't have flat plains like the front!"

"Look! The Moon's surface looks like a World War I battlefield!"

"Gemini VI and VII are almost touching!"

"Moon dirt looks just like my garden dirt, and the rocks just like those on Earth!"

"Look! Mars looks like the Moon!"

Perhaps the reader can imagine some of the excitement these pictures generated when they were first seen by those of us in the space program.

It wasn't all a series of exultant "Looks!" Disheartening failures had preceded the Ranger VI mission, but after a fine launch, smooth midcourse correction, and precise 2½-day voyage, all problems seemed to have been solved. Then in the last minutes before lunar impact, the cameras failed to turn on. Nor will we easily forget the thruster malfunction on Gemini VIII that almost brought tragedy. But perseverance prevailed and more than 90 percent of space missions attempted in 1967 were successful. Browning once wrote, "Ah, but a man's reach should exceed his grasp, or what's a heaven for."

This book is based on a lecture that I gave on November 16, 1966, at the annual banquet of the Washington Society of Engineers. The lecture was subsequently repeated many times, usually drawing advice to write a book based on the pictures. Instead of writing it myself, I decided to ask my friends and associates in the space program to help produce this book by preparing captions for pictures having special meaning for them. In some cases, the captioneers played a major role in the program that produced the pictures. In other cases, they are experts on the pictures' contents. They include scientists, engineers, and administrators. They responded enthusiastically, and I regret that some captions had to be dropped and others compressed in final picture editing. I hope all of the contributors enjoy *Exploring Space With a Camera.*

In this book I have used primarily photographs from the space program of the United States, in part because they were most familiar and available to me. Some Soviet photographs have been included, however, to highlight some of the many accomplishments of their space program.

Many aspects of space exploration cannot be depicted by a camera. A hint of this can be seen on page 181, which shows a spectrograph of the star Canopus. Scientific studies of the upper atmosphere, the ionosphere, and the particles and fields in the great radiation belts of the magnetosphere are made primarily with instruments other than the camera. Nevertheless, this photographic essay depicts the imagination and courage of space exploration in a way that should hold deep meaning for participant and spectator alike.

It is also an exciting portent for the future, when space flight will play a role of ever-increasing importance both for science and for the practical benefit of man. Our unearthly spacecraft will help us solve a host of earthly problems. They are already indispensable to global communications and will play a major role in breaking the communications barriers of the world. Dramatic improvements in all-weather navigation and in the traffic control of ships and aircraft are emerging. Satellite collection of meteorological data will make possible long-range forecasting of global weather. And with the aid of satellites, man will come to monitor and manage many of the natural resources of the Earth.

Soon now, man will tread the barren wastes of the Moon. One day, after suitable reconnaissance, Mars will yield its secrets in like manner. Space stations will perpetually circle the Earth in the conduct of both scientific observations and technological investigations. And precursor spacecraft will probe the outer regions of the solar system, and peer tentatively toward the starry, endless universe beyond.

EDGAR M. CORTRIGHT
Director, Langley Research Center
National Aeronautics and Space Administration

ABOVE

THE ATMOSPHERE

A satellite 22 300 miles above the Earth at 95° W longitude photographed the storms both north and south of the equator January 21, 1968. "The color camera worked well," JOSEPH R. BURKE, Applications Technology Satellite Program Manager, reported, "although it had been turned off for about a month to let unexpected gas accumulation leak out. The ground equipment still was being tuned to give a better color balance. Both coasts of both North and South America are visible, and you can glimpse the Great Lakes through the clouds over Canada."

2

ABOVE THE ATMOSPHERE

Never has the statement "It's a small world" seemed as true as since the advent of the artificial satellite. Less than a century ago circumnavigation of the Earth was measured in years. The steamship shrunk this time to months and then weeks. The aircraft reduced the time for such a journey to days.

The satellite has revolutionized all this. At low altitudes a satellite circles the globe in about 90 minutes, and by sequential passes makes possible detailed observations of the entire Earth and its atmosphere twice a day. From very high altitudes, such as the 22 300-mile geostationary orbit, a satellite can remain ever watchful of the scene below and in uninterrupted communication with nearly half the Earth. Thus man's ability to observe his dynamic environment has for the first time begun to equal his needs.

Meteorology has been one of the first beneficiaries of space flight. The Earth's atmosphere may be likened to a gigantic heat pump, driven by the energy from the Sun and the coriolis forces of the rotating Earth, and throttled by its own gigantic energy exchanges of evaporation and condensation. Weather everywhere is coupled to weather everywhere else, if not today—then tomorrow, or next week, or next month.

This dynamic global character of meteorology demands frequent if not continuous coverage on a global basis if man is to fully understand his atmospheric environment, predict it well in advance, and someday even influence it. Before the satellite most of the Earth's weather went unobserved since most of the Earth's surface is uninhabited. Today that has changed—thanks to the meteorological satellite.

The photos on the next two pages were taken before a camera could be kept in orbit. Such demonstrations of its usefulness in space prompted the suggestion that the difficulties of recovering films be obviated by placing television cameras on satellites. This was done with Tiros I in 1960, and transmissions from the Tiros satellites have been a mainstay for 8 years now in meteorological research. During the early excitement, Tiros was mentioned in the title of nearly every scientific paper in which information from it was used. Now the availability of such data is assumed.

Operational meteorologists, meanwhile, have expressed ever more strongly their desire for quantitative measurements such as the early researchers needed. The use of radiometers to take measurements in both visible and infrared regions has become routine.

To receive the first satellite telemetry carrying cloud picture data, multimillion-dollar command and data acquisition stations were needed. Sending it on to local forecasters via landlines or radio often delayed and degraded the quality of the pictures. Placing an Automatic Picture Transmission System on satellites has removed this difficulty. With relatively inexpensive receiving equipment, any station within the line of sight from a satellite that is automatically transmitting pictures now can receive a picture directly within 3 minutes of the time it was taken.

Addition of color channels to a camera of the Applications Technology Satellite (ATS) type has made it possible to obtain views such as the one on the opposite page and the frontispiece of this book. With ATS–III, color pictures can be taken every 24 minutes. When taken throughout a day, researchers can study both cloud motions over the entire disk and color changes in the day-to-night terminator as it moves across the Earth. These color changes are expected to be helpful in determining heights of clouds and possibly other matters of interest to meteorologists. The time-motion study of weather dynamics made possible by such synchronous satellites as ATS–III may add a new dimension to our understanding of atmospheric processes.

Prelude to Progress

Cameras carried on sounding rockets and missiles first demonstrated the value of high-altitude photography in meteorology. These pictures were recovered from film packs carried aloft. They stimulated the development and employment of equipment to televise views obtainable from satellites to stations on Earth. The photo below was taken from a development flight of a missile high over the Atlantic Ocean on August 24, 1959.

GEN. DONALD N. YATES, who commanded the Air Force Missile Test Center in the 1950's, recalls that: "A 16-mm motion picture camera was placed on the stabilization system during an Atlas flight. This photograph was taken early in the flight, showing the booster separating from the nose cone, with the Earth's horizon reflected from its polished surface."

From less comprehensive pictures than the one on this page, mosaics such as those on the next page were put together. The upper one consists of photos taken in 1954 by an Aerobee sounding rocket that was launched from White Sands, N. Mex.

LESTER F. HUBERT, of the National Environmental Satellite Center, remembers it well: "The spiral cloud pattern in the upper left center was produced by a tropical disturbance that had moved over Texas from the Gulf of Mexico. Decreasing in intensity, it did not disturb the surface winds, but maintained a tight cyclonic circulation in the upper atmosphere. At a glance, this picture documented a flood-producing circulation that was undetected by routine means after moving over Texas.

"Photography from rockets stimulated the meteorological satellite program, but no previous picture had been as dramatic as this, nor as convincing that satellite surveillance of hurricanes was feasible. Here was the first demonstration that storms could be detected by ultra-high-level photographs (since borne out by thousands of satellite pictures).

"This mosaic was constructed by Otto Berg, who was in charge of the Naval Research Laboratory's rocket photography."

The lower picture on the next page was a further demonstration of the potentialities of space technology. GEN. BERNARD A. SCHRIEVER, USAF (Ret.), explains how it was obtained:

"Taken from an altitude of 700 miles, this panoramic photograph was an early spectacular achievement of our space effort. It shows an Earth segment from North America through the Caribbean to the coast of Africa, including massive storm fronts over the Atlantic Ocean and the Tropics.

"The panorama was constructed from motion pictures taken by a 16-mm camera carried on the Mark 2 reentry vehicle of our Atlas 11C R&D flight of August 24, 1959. This flight was one of a series in which we experimentally stabilized the reentry vehicle with an infrared horizontal stabilization system. The camera was in the capsule that was ejected from the reentry vehicle and recovered."

4

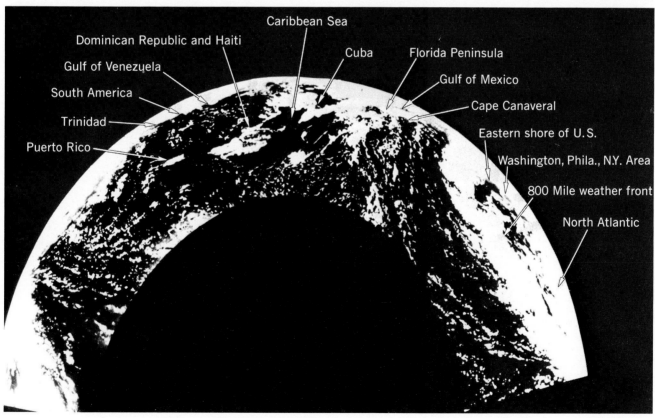

Caribbean Sea

Dominican Republic and Haiti

Gulf of Venezuela

Cuba

Florida Peninsula

South America

Gulf of Mexico

Trinidad

Cape Canaveral

Puerto Rico

Eastern shore of U.S.

Washington, Phila., N.Y. Area

800 Mile weather front

North Atlantic

The Clouds
Draw a Map

The first Tiros (Television Infrared Observational Satellite) began observations on April 1, 1960, that have figured prominently in meteorological progress ever since. It transmitted 22 592 pictures to Earth during its 1302 revolutions. The mosaic at the right illustrates the way many of those pictures were assembled and used.

"From the very first," says MORRIS TEPPER, Deputy Director, Space Applications Program, NASA, "pictures taken by the Tiros satellites showed that the Earth's cloud cover was highly organized on a global scale. Coherent cloud systems were found to extend over thousands of miles and were related to other systems of similar dimensions. Moreover, it was soon readily apparent that these cloud patterns were in fact 'signatures' of weather systems.

"The upper portion of the figure is a mosaic of overlapping video pictures taken by Tiros I on May 19 and 20, 1960. Below it is the associated weather map showing three active storm systems: from left to right, a very intense one in the North Pacific, one on the west coast of the United States, and one in the Midwest. The cloud structure as seen by Tiros has been superimposed on the weather map. It is remarkable how closely the cloud systems delineate the weather systems. It is as if Nature were actually drawing her own weather map directly onto the Earth."

Many more such mosaics were produced as the Tiros flights continued. The most significant milestone passed in these flights, says ROBERT M. RADOS, NASA's Tiros Project Manager, was "to demonstrate the feasibility of an operational meteorological satellite system and its application to regular worldwide weather analysis and forecasts, aimed toward increasing man's ability to understand and cope with his physical environment."

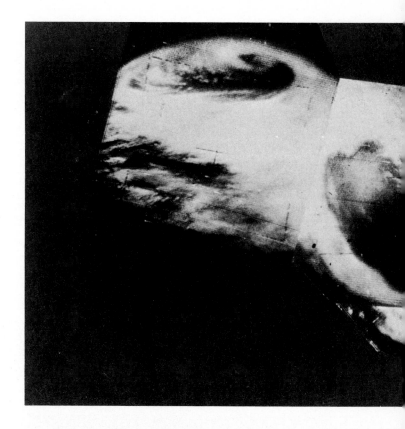

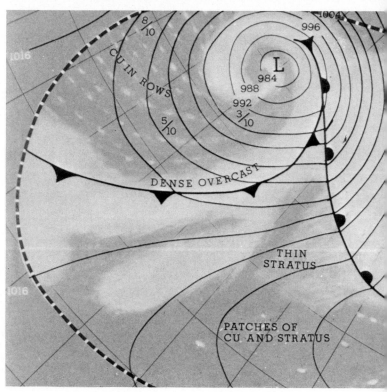

TIROS I

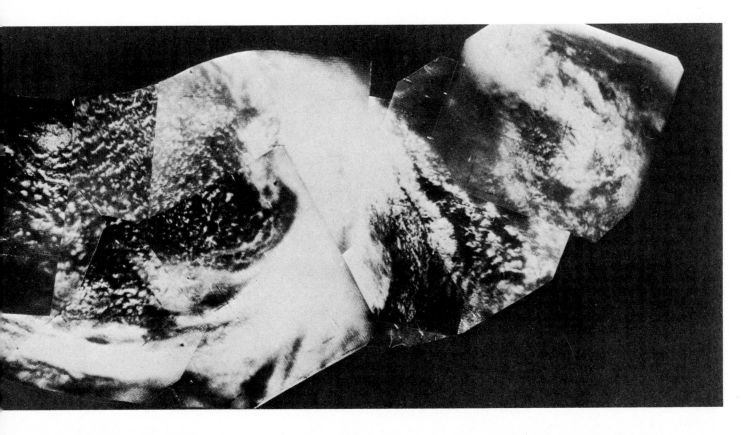

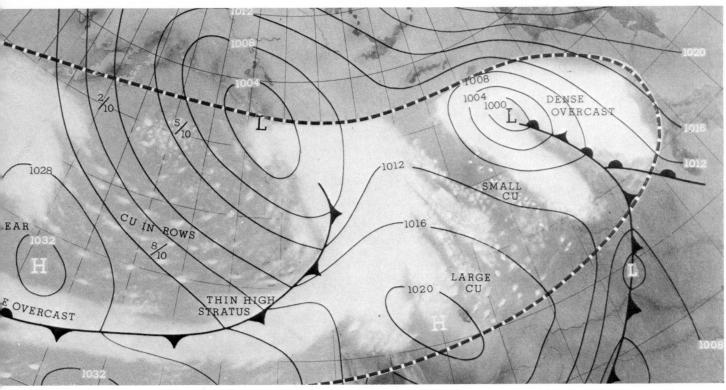

2/10

5/10

1028

8/10

CLEAR

1032

H

CU IN ROWS

E OVERCAST

1032

THIN HIGH
STRATUS

L

1012

1008

1004

1012

L

1016

1020

H

LARGE
CU

1000

1004

1008

L

DENSE
OVERCAST

SMALL
CU

1020

1016

1012

1008

L

A Satellite Discovers a Hurricane

... and Begins to Save Lives

"Sometimes," William Shakespeare noted, "we see a cloud that's dragonish"—but not always soon enough from Earth's surface. Hurricanes emerge from large bodies of warm, moist air near the equator, vary their courses, and differ in ferocity. Tiros III photographed 1961's first five hurricanes from an altitude of more than 400 miles—and that's how Esther, the big one on the next page, was discovered.

S. FRED SINGER, Deputy Assistant Secretary for Scientific Programs, Department of the Interior, provided this account of the hurricanes pictured below and the dragonish sight on the facing page:

"Esther was the first hurricane to be discovered by a satellite, but Tiros III also took valuable pictures of Anna, Betsy, Carla, and Debbie in 1961. In all cases, the Tiros pictures provided important supplements to reconnaissance aircraft observations and to data obtained by ground-based observations, by fixing the position of the hurricane center and showing the extent of the spiral cloud bands.

"The photographs here were selected to show all of the hurricanes at the stage of their maximum development; their organization in spiral patterns can be seen very clearly. It has now become possible to deduce the strength of the hurricane winds from the degree of organization seen in satellite pictures.

"Hurricane Anna went into the southern Caribbean and hit Central America.

"Betsy, on the other hand, never was a threat to land areas and apparently caused no damage to shipping.

"While Betsy was developing in the Atlantic, however, Carla—a giant hurricane—was discovered some 300 miles north of Panama. It eventually turned out to be one of the most destructive hurricanes in recent history, causing 46 fatalities in the gulf coast region and over $300 million property damage in Texas and Louisiana. More than 300 000 people were evacuated from the gulf coast.

"Debbie never hit North America, but curved back and caused deaths and heavy damage on the western coasts of Ireland and Scotland.

"Esther will be assured a place in meteorological history. She was the first hurricane to be discovered by satellite, moved in a complex loop off the east coast of the United States, threatening New England, and was the guinea pig for a seeding experiment with silver iodide."

Anna

Betsy

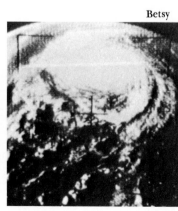

Carla

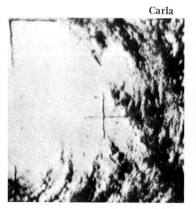

Debbie

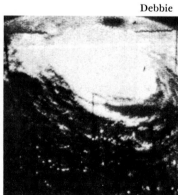

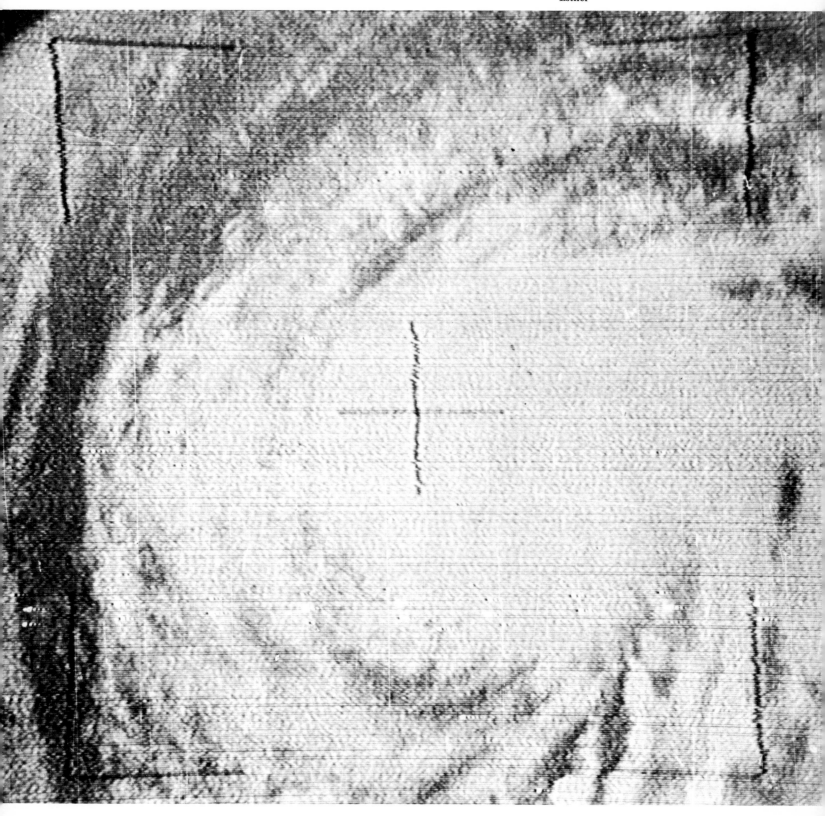

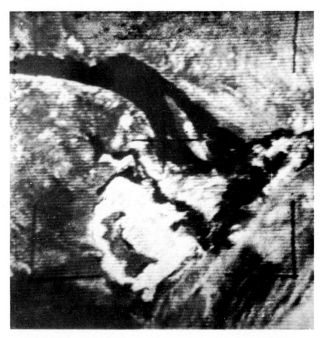

April 3, 1962

April 5, 1962

Ice Is Tracked . . .

Four photos at the tops of these pages showed satellites can alert seamen to ice as well as to storms. J. R. H. NOBLE, Director, Meteorological Branch, Department of Transport, Canada, pointed out the significance of these photographs:

"Cameras of Tiros IV focus on the Gulf of St. Lawrence. The river estuary and northern portion of the gulf contrast sharply with the dull gray of land. The ice retreating southward shows brilliantly white, and clouds are reproduced in a variety of shades. In the first picture the icepack surrounds Prince Edward Island, stretching toward the Magdalens with ice tongues from Chaleur Bay and southward from the Strait of Belle Isle. Following the sequence through April 12, the main pack is gradually driven southeastward. On the last day, ice areas appreciably decrease with the drift past Cape Breton Island. An interesting feature is the amount of ice held back by Cape Breton and Magdalen Islands. Information from visual sorties for ice reconnaissance along the ice edges confirmed to a remarkable degree the results from the satellite; thus space technology now assists mariners."

and a Cold Line Opens

ARNOLD FRUTKIN, Assistant Administrator for International Affairs, NASA, reports regarding the picture at bottom right:

"In June 1962 representatives of NASA and of the Academy of Sciences of the U.S.S.R. agreed upon a cooperative meteorological program. The two sides were to develop first experimental and then operational meteorological satellites capable of photographing the Earth's cloud cover. They were to establish a conventional high-capacity communications link between Washington and Moscow, dividing the cost, to exchange cloud photographs obtained by their respective satellites, together with analysis of cloud information.

"By November 1964, the communications link called the 'cold line' had been established. As 1967 drew to a close, the link had been used to exchange conventional weather data and a few satellite cloud-cover photographs per day. It must, therefore, be regarded as in the earliest experimental stage. The accompanying photograph is an example of the cloud-cover photographs provided by the Soviet Union over the 'cold line.'"

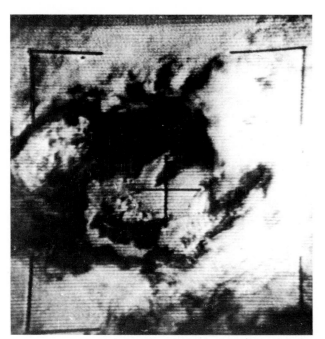

April 7, 1962

April 12, 1962

480 Pictures in 24 Hours Girdle the Globe

"Tiros IX provided this dramatic 'first' complete view of the world's weather," according to ROBERT M. WHITE, Administrator, Environmental Science Services Administration. "As Tiros IX circled the Earth once every 2 hours in its polar orbit, its 2 television cameras obtained 40 pictures on the Sun-illuminated side of the Earth. Once each orbit the pic-

ture signals were received by radio at ground stations in Alaska and Virginia and immediately relayed to Washington, D.C., where the signals were converted into picture form. After circling the Earth 12 times during the 24-hour period of February 13, 1965, all of the Sun-illuminated portion of the Earth had come within the field of view of the cameras aboard the satellite. The 480 pictures taken during this period were then placed together to produce this bird's-eye view of the world's cloud systems.

"A tropical storm can be seen over Ceylon and the southern tip of India, and another is over the south Indian Ocean. In the lower right, a storm is ap-

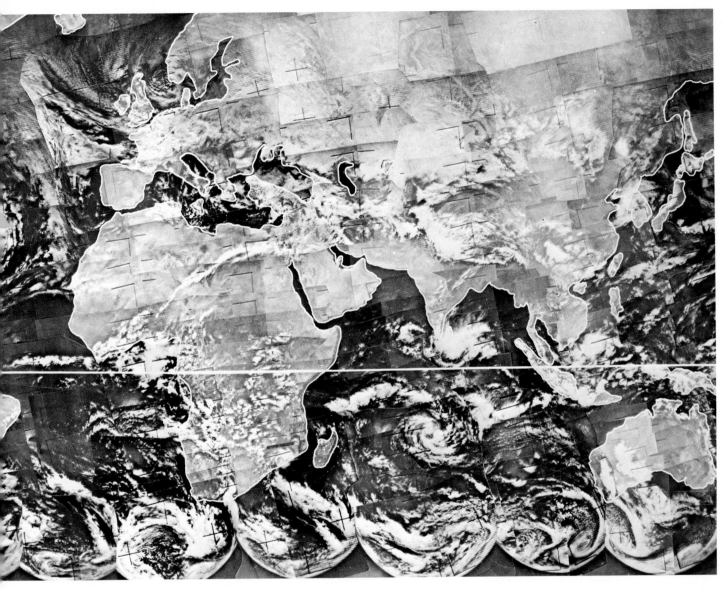

proaching the southern coast of Australia; this storm had moved approximately 500 miles to the east 24 hours later, as seen in the pictures of the same area at the left of the mosaic. The thin band of clouds extending from central North Africa across the Red Sea to Saudi Arabia indicates the location of the jet stream in this part of the world, a strong current of air moving from west to east high above the Earth's surface. The remnants of an old storm are indicated by the comma-shaped cloud array over the North Atlantic Ocean.

"A strong weather front is depicted by the clouds extending across the southeastern United States; an-other storm is moving into the northwestern United States from Canada.

"The importance of this picture lies in the fact that it provides the meteorologist with weather information over the entire Earth, whereas conventional observations before satellites provided information on less than 20 percent of the Earth's atmosphere. This global observing capability of space platforms, and the rapid development of space technology, has led to the establishment by the United States of the world's first operational weather satellite system, which is now providing similar pictures of the world's weather every day."

Latest Pictures
Are Broadcast

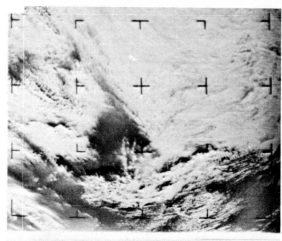

Automatic Picture Transmission (called APT) that was flown on Nimbus I reduced the time required and the cost of getting photographic information from a satellite to local weather forecasters. The four pictures at the right were received at Wallops Island, Va.

F. W. Reichelderfer, former Chief of the U.S. Weather Bureau, wrote of them: "In this four-frame sequence, Nimbus I reveals many things about the weather in a 900-mile swath from Venezuela to the Canadian Arctic as viewed on August 29, 1964, from orbit approximately 550 statute miles from Earth. Through APT, the local or regional weather forecaster receives pictures of the cloud arrays as the satellite passes overhead. APT receivers are fairly simple and inexpensive.

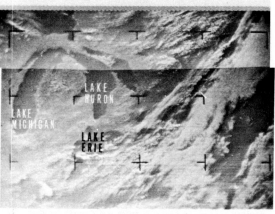

"The two top pictures show clouds typical of systems of fronts. In midlatitudes and the subarctic, warm fronts produce cloud 'decks' of cirrostratus and altostratus. Cold fronts produce convective cumulus and line-squall cumulonimbus. Intermediate frontal forms give the complex combinations of cloud forms and arrays seen in the pictures—their orientations and shapes being indicative of different wind currents, etc.

"The third picture looks down upon historic Hurricane Cleo over South Carolina and Georgia. The characteristic spiral bands of clouds first seen in full by rocket and satellite photos make it easy for satellites to discover tropical cyclones far at sea outside the usual networks of ship reports.

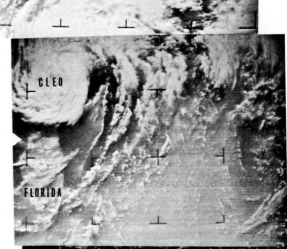

"In the fourth picture the many cumulus 'streets' often found in the trade winds can be seen; also southward are the cloud masses of an intertropical convergence zone.

"Perhaps most promising for future weather analysis and forecasting is the increasing study of clouds and their arrays as symptoms, and thus diagnostic means for identifying atmospheric dynamic systems which make the weather."

14

In-Depth
Research

"As Hurricane Gladys neared the southeast coast of the United States (66.5° W, 25.5° N), the high-resolution infrared radiometer (HRIR) scanning system aboard the Nimbus satellite acquired the facsimile, at the right, of the clouds of the hurricane and the surrounding water masses," according to WILLIAM G. STROUD, of the Goddard Space Flight Center, NASA.

"At the time, Gladys was a fully mature hurricane about 450 miles in diameter. Winds were about 125 mph near the center, with hurricane winds extending 115 miles northward and 85 miles southward. Nimbus I tracked Gladys through its 12-day life as a hurricane.

"The HRIR aboard Nimbus I and II senses radiation from the land, water, and cloud surfaces in the 3.6- to 4.2-micron region of the spectrum. The shades of gray are proportional to the blackbody temperature of the radiating surfaces; the white areas are the tops of the clouds and the coldest areas; the black areas are the surface of the sea, and warmest.

"Below the picture a single scan line passing through the center of the hurricane is displayed. On the left is the nonlinear equivalent-blackbody temperature scale. The scan line is from horizon to horizon, about 5000 km along the surface of the Earth, highly distorted at the edges. At the western (left) horizon the temperatures rise sharply from 210° K of the sky to 290° to 300° K, the approximate sea-surface temperatures. The temperatures at the cloud tops are as low as 210° to 220° K, and the observed temperatures in the eye of Gladys are about 290° K. These temperatures are converted to height above sea level by equating them to actual temperatures measured by sounding balloons in the vicinity of the storm, as shown on the right. The 290°-K temperature over the eye of the hurricane corresponds to a height of about 2 km; the radiometer

probably did not see the surface of the sea through the eye because of the usual high-level, thin, cirrus clouds covering the eye. The main cloud tops were at a height of about 12 km."

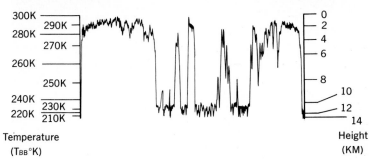

Night Temperatures and Clouds

Nimbus I began a second generation of meteorological satellites in 1964. The Tiros satellites each had two cameras; Nimbus had three. It also bore other advanced equipment to be tested for use in future weather-satellite systems. Since the clouds are "birds that never sleep," the equipment included devices for collecting information about them over dark as well as sunlit areas of the Earth.

SIGMUND FRITZ, Director of the Meteorological Satellite Laboratory of the Environmental Science Services Administration, noted that the picture at the left illustrated one of the additional services to his profession provided by Nimbus I. He writes:

"This picture was made from quantitative measurements taken with the Nimbus I high-resolution infrared radiometer on September 12, 1964, at 2340 GMT. The area extends from Scandinavia through Italy to the Tunisian coast. In the picture, dark areas represent warm sources, and bright areas denote cold sources. Thus the colder land areas appear brighter than the surrounding water. Moreover, the very bright areas represent cold cloud-top systems whose heights can be estimated from the measurements.

"The meteorological situation can be to a certain extent deduced from the picture. The cloudless area over northwestern Europe was associated with an extensive high-pressure area. The clouds were mainly associated with low-pressure areas and frontal systems. Thus these Nimbus data can locate significant cloud systems at night, give insight into meteorological situations, and aid in estimating the heights of cloud tops."

"This is a Nimbus II high-resolution infrared radiometer picture taken near local midnight on October 7, 1966," HARRY PRESS, Nimbus Project Manager, Goddard Space Flight Center, NASA, says of the picture on this page. "It covers a wide swath from 15° N to 50° N over the eastern United States and Caribbean.

"The spiraling cloud mass at the bottom is Hurricane Inez whose center was located just north of the Yucatan Peninsula. The Great Lakes are visible at the top of the picture. Clearly distinguishable is the eastern coast of the United States from Maine to Georgia.

"Temperatures measured by the infrared scanner are converted to shades of gray, from white to black: the white being the coldest and the black the warmest temperatures measured. Grid points are at every two degrees of latitude and longitude.

"The white band of clouds off the east coast is associated with a cold front over the Atlantic. Noticeable are the colder water temperatures near the coast north of Cape Hatteras. The ocean here appears lighter (colder) than the dark (warmer) clear areas in the Caribbean and south of Cape Hatteras. The warmer temperatures south of Cape Hatteras and near the frontal cloud band outline the warm waters of the Gulf Stream."

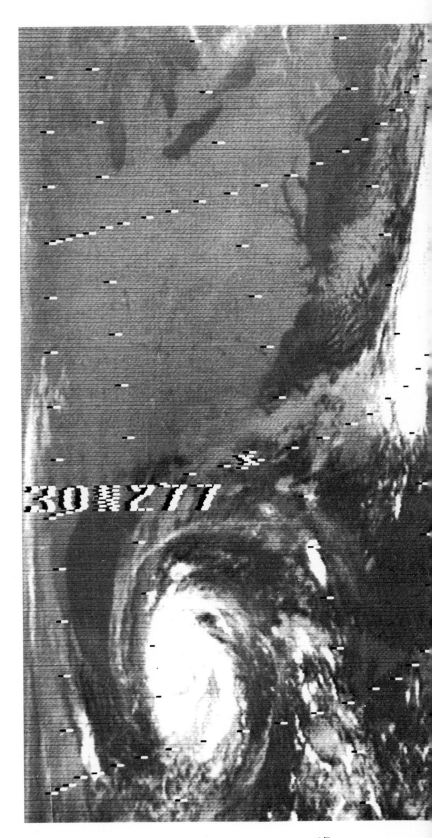

Europe
Tunes In

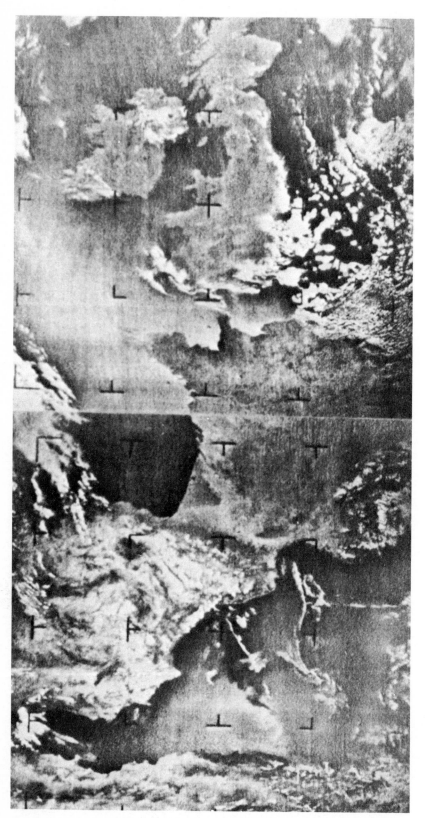

In Lannion, France, on August 31, 1964, meteorologists received the pictures at the left from a passing Nimbus, showing western Europe from Scotland south to the Mediterranean.

JEAN BESSEMOULIN, Chief Engineer of Meteorology for the French General Secretariat of Civil Aviation, wrote of this display:

"On this assemblage of two successive photographs taken by satellite Nimbus I during its 46th orbit, France appears in the center exceptionally well, due to a sky almost completely devoid of clouds. The mirrored reflection of the Sun on the Atlantic and on the English Channel accentuates the clearness of the coastal outline of Brittany and Normandy. The Alps and the Pyrenees show some snow-covered peaks. In the southwest is an outline of the triangle of the forest of Landes. The British Isles also enjoy a generally clear sky, with the exception of cumulus clouds visible particularly on the south coast and on the coast east of England.

"In the Mediterranean, the Balearic Islands, Corsica, and Sardinia are marked by clouds of diurnal evolution. And finally, to the northwest of the Iberian Peninsula, appear stratified cloud formations, in connection with perturbances located on the open sea in the Atlantic.

"The Lannion Center of Meteorological Space Studies exploits constantly this type of photograph, from which it establishes maps of cloud formations which are disseminated to meteorological centers on national and international channels."

CHARLES J. ROBINOVE, of the U.S. Geological Survey, commented as follows on the picture at upper right:

"The lower Nile-Sinai Peninsula picture taken by the Advanced Vidicon Camera System (AVCS) shows the open water and irrigated areas (dark), and the rocky and sandy desert (white and gray), and some of the high-contrast drainage patterns. Terrain features of high contrast show well in the picture, but low-contrast features do not. The dark area at left center is the El Faiyum depression, in which crops are irrigated by water diverted through canals from the Nile River. Although the AVCS camera is designed for meteorological use, striking and useful pictures of the Earth's surface have been taken of cloud-free areas. The ability of such space-acquired imagery to portray large areas of the Earth's surface at a single instant of time is of great advantage to the Earth scientist in assessing and mapping large regional features and their relationships."

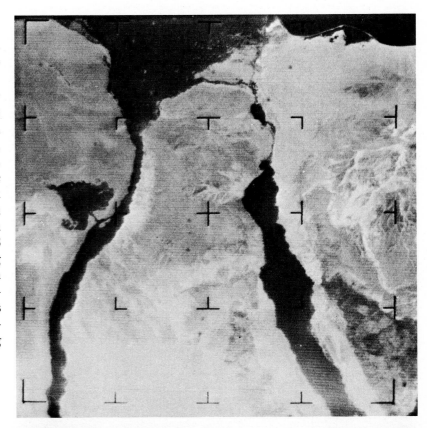

The lower photo here is a high-resolution picture of part of the same area shown on page 18 in lower resolution. STANLEY WEILAND, Nimbus Observatory Systems Manager, Goddard Space Flight Center, NASA, points out how much it encompassed: "The photograph comprises an area of about 41 600 nautical square miles from a height of 409 miles. The Pyrenees show as a dark area across the center. Mountains usually show darker than lowlands because of their thicker vegetation cover and basalt rocks. Another group of mountains, the Montagne Noire, appear as a dark tongue near the center right edge. The valleys of the Ebro River and its tributaries in western Spain (lower left quadrant) are a good example of a flood plain. Also evident because of its dark soil is the Grande Landes, an alluvial rich-humus flood plain in southwestern France (upper left quadrant). The course of the Garonne River can be traced to the junction with its tributary the Grou River, just north of Toulouse (center and upper right quadrant)."

AVCS

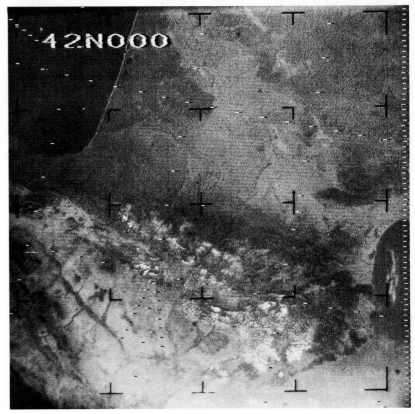

More Than Men See

Multispectral imaging can tell scientists more about the cocoon of air around the Earth than human eyes can see. It is simply a matter of combining observations made with instruments that respond to different wavelengths in the electromagnetic spectrum — a technique nicely illustrated here.

WILLIAM NORDBERG of the Goddard Space Flight Center, NASA, explains this figure this way:

"Pictorial presentations of radiation emitted and reflected by the Earth and the atmosphere in various spectral bands have radically improved our capability to observe meteorological features from satellites. When interpreted quantitatively, these measurements lead to the description of the telluric temperature, cloud, and moisture fields. Results shown here were obtained with Nimbus II on May 17, 1966.

"Measurements were made in the five spectral regions of: total reflected solar radiation (0.2 to 4.0 microns), total emitted thermal radiation (5 to 30 microns), CO_2 absorption (14 to 16 microns), the atmospheric "window" (10 to 11 microns), and water vapor absorption (6.4 to 6.9 microns) as Nimbus II passed from the south-central United States (bottom) over the Pacific, Antarctica, the Indian Ocean (center), India, and Siberia to North America. "The radiative energy balance can be determined from the difference between the first two channels. The CO_2 channel maps out the temperature field in the lower stratosphere, the water-vapor channel gives the moisture content in the upper troposphere, and the window channel shows Earth-surface temperatures and cloud heights. The hot surface, and moisture-laden atmosphere, of India are obvious in the picture (upper third). In emission, low-radiation intensities (cold, cloudy, or moist) are shown light; high-radiation intensities (warm, clear, dry) are dark. The reverse is depicted by reflection (0.2 to 4.0 microns)."

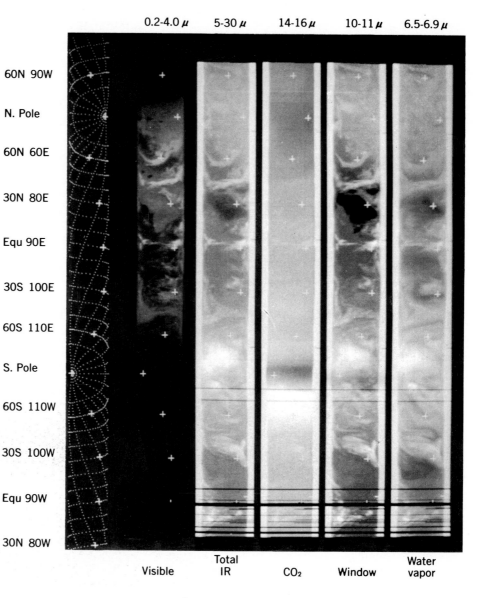

0.2-4.0 μ   5-30 μ   14-16 μ   10-11 μ   6.5-6.9 μ

60N 90W
N. Pole
60N 60E
30N 80E
Equ 90E
30S 100E
60S 110E
S. Pole
60S 110W
30S 100W
Equ 90W
30N 80W

Visible — Total IR — CO_2 — Window — Water vapor

The Curls
in Clouds

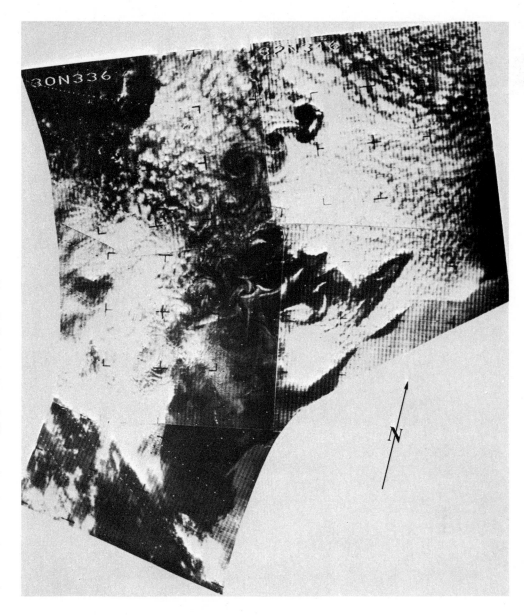

For poets the clouds have long been "a volume full of wisdom." Now it is easier to read. The curls at the right are an example of the details of the turbulent atmosphere that formerly eluded the eyes of both artists and scientists.

GUENTER WARNECKE, of the Planetary Radiation Branch of the Goddard Space Flight Center, NASA, points out:

"This picture is a mosaic of photographs taken by Nimbus II off the west coast of North Africa at noontime on July 16, 1966. The Moroccan coast is recognizable under clear sky conditions; Cape Blanco is the prominent landmark in the lower section of the mosaic. The bright pattern exhibits a low-level cloud system embedded in the stable and permanent north-northeasterly air flow of the trade-wind regime of the subtropical Atlantic Ocean.

"The appearance of vortex structures within the cloud deck was not visually observed in the atmosphere before satellites became an observational tool, as these vortex patterns are too large to be observed from the surface and too small to be resolved in conventional weather maps.

"The northernmost series of the vortices is of outstanding regularity over a distance of more than 400 miles and reflects the apparent existence of a Kármán vortex street underneath the cloud level in the wake of the 6100-foot-high obstacle of the island of Madeira. Farther south the irregular group of the Canary Islands creates a more complicated assembly of curls and horizontal wavelike structures reflecting the superposition of Kármán-type vortex streets and gravity waves induced on the low-level trade-wind inversion by the mountainous islands."

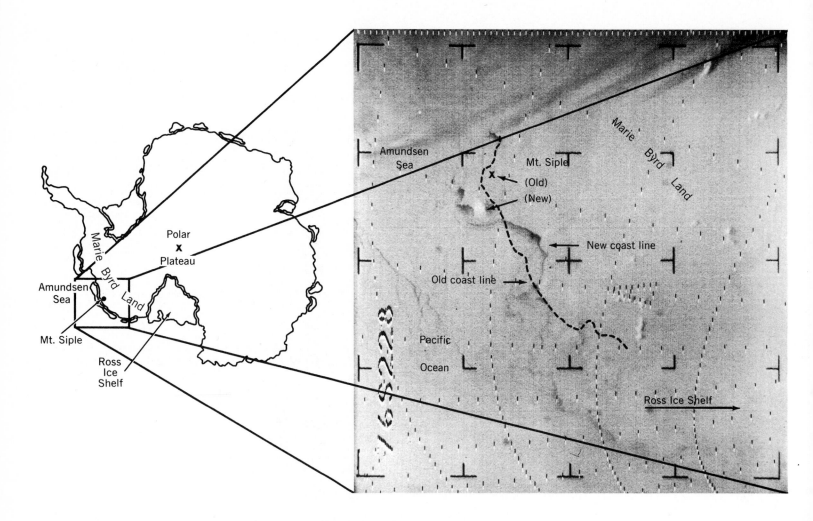

Help for Cartographers

RICHARD L. HALEY, Nimbus Program Manager, Space Applications, NASA, wrote of this picture taken September 9, 1964, from high over Antarctica: "In addition to the use of Nimbus data for meteorological purposes, the data have been utilized in such scientific disciplines as glaciology, geology, and oceanography.

"For example, here is a Nimbus I Advanced Video Camera System (AVCS) picture taken as the satellite passed 950 kilometers above the ice-covered Antarctic Continent. The picture covers an area 530 by 530 kilometers approximately in the Marie Byrd Land and Ellsworth Highland region.

"Just to the left of center is the Getz Ice Shelf. The dark areas delineating the coast are probably

due to pressure ripples and faults in the ice which attenuate the reflectivity of the surface. Cirrus cloud bands and their associated shadows produce the striations at the top of the picture.

"The Nimbus pictures have been extremely valuable in the mapping of the little-known Antarctic Continent. This and other pictures aided cartographers to relocate Mount Siple correctly 2° farther west on their maps. Mount Siple is visible as a gray-white spot just above the Getz Ice Shelf. Other mountain ranges in Antarctica were correctly relocated, and the configuration of the ice fronts in the Filchner Ice Shelf, Weddell Sea, and Princess Martha Coast areas were determined more accurately and completely than ever before."

Every Day's Weather Everywhere
Becomes a Matter of Record

DAVID S. JOHNSON, Director of the National Environmental Satellite Center of the Environmental Science Services Administration, explains this picture's significance:

"This satellite view of Europe and North Africa obtained by ESSA I from 478 miles above the Earth's surface represents in capsule form the culmination of 6 years of space development by the United States, resulting in the establishment of the world's first operational weather satellite system. Here the meteorologist has a picture of weather conditions over an entire continent, plus sizable portions of two others. Conditions ranging from snow-covered Scandinavia and ice-covered Gulf of Bothnia in the north to the cloud-free area of North Africa are discernible at a glance. The exact location of a low-pressure center over Denmark is unmistakably outlined by its characteristic circular array of clouds.

"The value of this mosaic is intrinsically great. On the day it was taken, March 1, 1966, it was used for pinpointing various meteorological features; today it, and thousands of similar pictures, are used by scientists in studying the Earth's atmosphere to increase man's ability to forecast the weather.

"The not-so-obvious value lies in the knowledge that this same area, indeed the entire Earth, was photographed the day before these pictures were taken, was photographed on all the succeeding days, and will continue to be photographed every day in the future. This is the essence of an operational weather satellite system: observation of the entire Earth reliably and regularly, day in and day out, for the benefit of man."

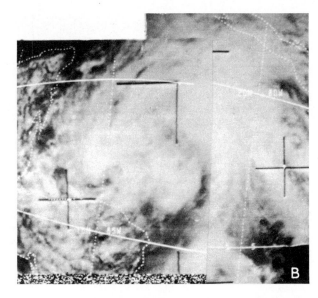

Alma's 9-Day Whirl

ESSA I photographed Hurricane Alma while she was at sea, shortly after noon each day from June 4 to 12, 1966, for meteorologists to chart her pressure.

L. F. Hubert of the National Environmental Satellite Center captioned these photos: "A grid of latitude-longitude lines at 5° intervals and dotted coastlines show that the storm became a hurricane just east of Central America and moved northward across Florida. The cloud masses in photos A and B reveal the intensification of a tropical storm in the Caribbean; C through H display patterns that are typical of mature hurricanes. These vast whirlpools of air are characterized by a calm, low-pressure eye which is clearly seen as a dark hole in the clouds in G and H.

"Hurricane intensity is inversely proportional to its central pressure; the graph illustrates the fluctuating pressure of Alma. Photo C was taken a few hours after Alma had reached hurricane intensity; the windspeed was 75 knots. On June 9, photo E,

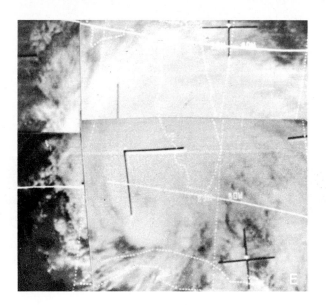

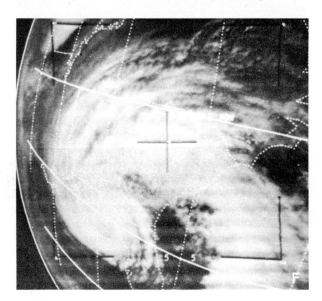

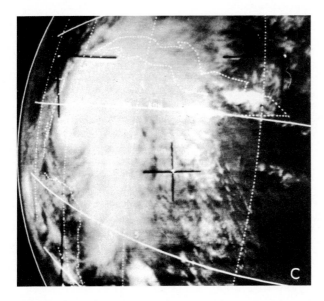

C

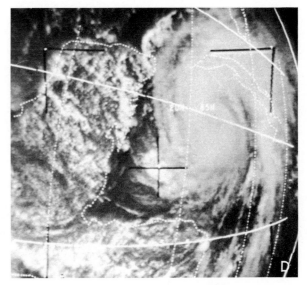

D

the maximum windspeed was about 110 knots, but the effect of the peninsula of Florida reduced the surface winds to about 95 knots on June 9, photo F. Traversing the peninsula of Florida, Alma regained some of her intensity on June 11, photo G, but declined soon after June 12, photo H.

"A technique has been devised to make reasonably accurate estimates of the maximum windspeed in storms by classification and measurement of their cloud patterns. Since February 1965, meteorological satellites have enabled meteorologists to classify and track every dangerous tropical storm, hurricane, and typhoon."

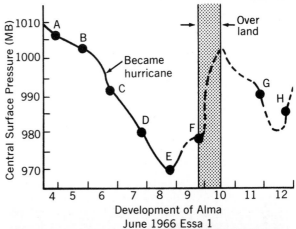

Development of Alma
June 1966 Essa 1

G

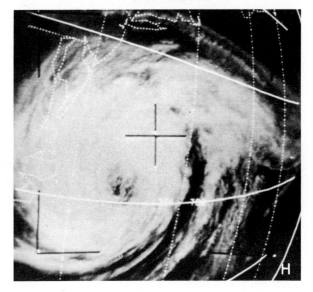

H

Weather Pictures
for the Asking

The Secretary General of the World Meteorological Organization in Geneva, Switzerland, D. A. DAVIES, wrote of the mosaic you see here:

"The launching of ESSA II added the Automatic Picture Transmission (APT) element to the world's first operational satellite system. Prior APT systems carried by Nimbus I and Tiros VIII had confirmed the enormous value of the facility for direct satellite readout at any point on the Earth's surface which the APT system presents. At the time of launching ESSA II, APT reception equipment had been installed at about 80 stations in 20 countries. Since then these numbers have increased steadily.

"This picture is a composite of 17 individual APT photographs taken on five orbits of ESSA II. The easternmost three orbits were acquired at Washington, D.C., and show how, within a space of some 4 hours, a clear presentation of the cloud systems over an area greater than the whole of North America may be directly obtained. The two westernmost orbits were obtained at San Francisco and Honolulu. The cloud formations of the depressions over the central Atlantic, near the Great Lakes, and off the west coast of the United States are very well displayed and follow closely the patterns to be expected from classical frontal theory. Indeed one of the revealing features of satellite cloud pictures is the striking confirmation they give of the frontal theory of the formation and development of depressions first enunciated by the Norwegian school of meteorologists nearly 50 years ago. The cloud formations of Hurricane Alma are visible in the western Caribbean Sea.

"The picture demonstrates also that cloudless land and water areas can be readily detected, including areas of frozen water in northern Canada, snow cover on the Rocky Mountains, and the icecap of Greenland."

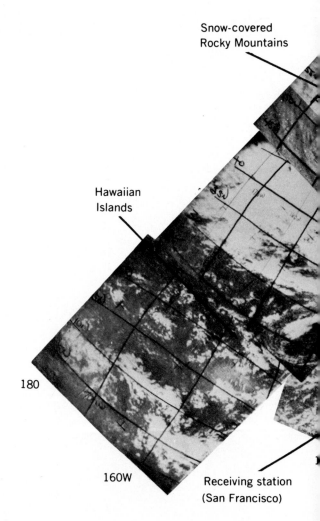

Snow-covered Rocky Mountains

Hawaiian Islands

180

160W

Receiving station (San Francisco)

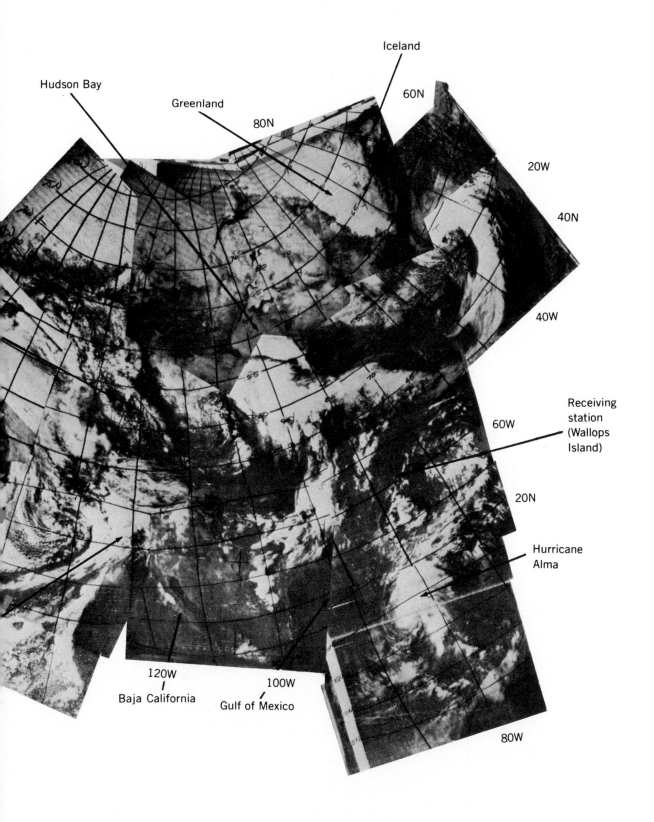

Hudson Bay

Greenland

Iceland

80N

60N

20W

40N

40W

60W

Receiving station (Wallops Island)

20N

Hurricane Alma

120W

100W

Baja California

Gulf of Mexico

80W

ESSA II

Characteristics of Churning Air

RICHARD J. REED, of the Department of Atmospheric Sciences at the University of Washington, welcomed pictures such as the one below because, in his words:

"Thin layers of fluid heated from below in laboratory experiments may, under proper conditions, overturn in cellular patterns. If the upper part of the layer is in motion relative to the bottom, the overturning may take place in long, horizontal rolls whose axes point in the direction of the shearing motion.

"Cloud photographs taken from satellites reveal that analogous cellular patterns occur in the lower layers of the atmosphere when frigid air crosses a warm surface. In this picture cold, dry air is streaming southward from eastern Alaska and western Canada over the relatively warm waters of the Pacific.

Moisture supplied by the ocean causes clouds to form in the cold-air stream after it has traveled some distance offshore. The initial cloud pattern consists of small cells arranged in bands, indicating that the heating is confined to a shallow layer of strong vertical wind shear. Farther south, where the heating extends through a greater depth, the cells are larger and have a honeycomb appearance. Rings of ascending, cloud-laden air surround open spots in which the air subsides.

"Pictures like this provide fundamental knowledge of the manner in which heat and moisture are transported from the ocean to the atmosphere. This knowledge will lead eventually to increased accuracy in long-range prediction where energy exchanges with the surface are an important factor."

MICHAEL L. GARBACZ, Tiros/TOS Project Manager, Space Applications, NASA, points out that the photo of Hurricane Faith at the right was used both to locate it, about 300 miles off Cape Hatteras on September 1, 1966, and to determine its important characteristics. "The general cloud organization," he notes, "reveals the windspeeds at the center were on the order of 110 knots. This picture and others like it on preceding and succeeding days provided local forecasters with information on which to base forecasts and warnings. The bright spot to the left is the reflection of the Sun on the ocean's surface."

LEE M. MACE, of the Environmental Science Services Administration, notes that photos such as the one at the right here not only expose frontal systems but also provide valuable information on the location of snow and ice to hydrometeorological services.

"Most of Scandinavia, under generally free cloud conditions, is clearly visible between two weather systems here," he points out. "The band of frontal cloudiness at the upper right has moved from west to east across Scandinavia, while another in the lower left is approaching. Abundant snow cover can be seen on the mountains of Norway and Sweden. The sharp valleys of the fiords are made visible by the apparent lack of snow and the shadows cast by mountains. Numerous frozen lakes with a snow layer on the ice are seen as bright patches in Finland. An area of solid ice covers the northern portion of the Gulf of Bothnia except for a wide northeast-southwest lead in the ice along the west coast of Finland.

"The peninsula and islands of Denmark appear in the lower center of the picture. Other islands in the Baltic Sea are outlined by sun glint on the sea.

"The appearance of the ice which is filling the northern Gulf of Bothnia and the pack ice northwest of Norway are of interest to those engaged in fishing and shipping."

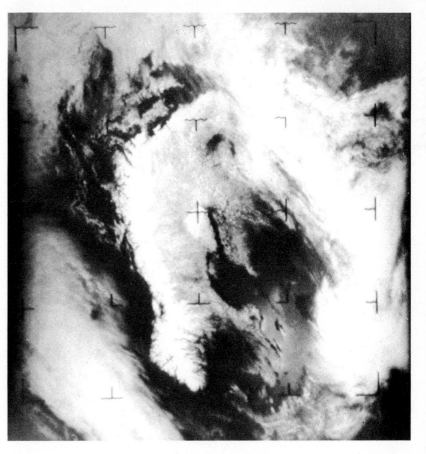

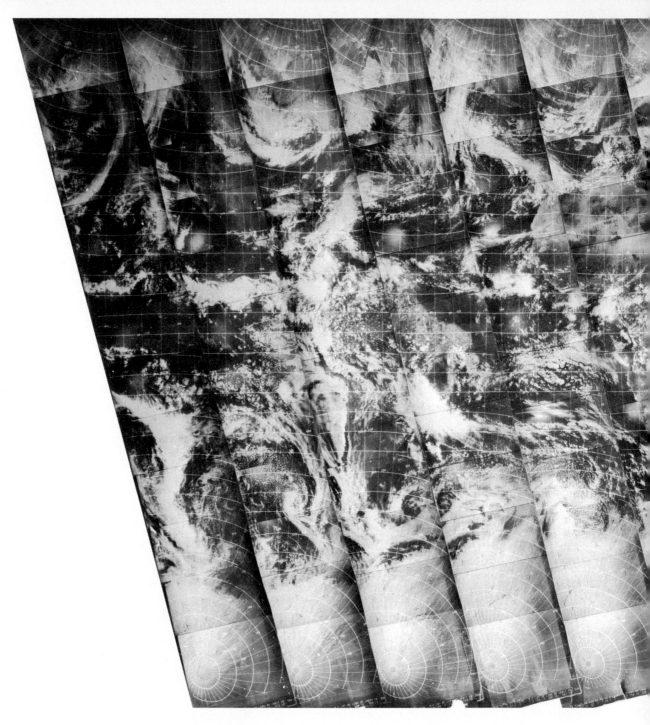

Daily Global
Weather Coverage

"This photomosaic depicts complete global coverage assembled from pictures taken during 12 consecutive orbits on October 31, 1966, by the ESSA III meteorological satellite," HERBERT I. BUTLER, Chief, Operational Satellites Office, Goddard Space Flight Center, NASA, tells us. "The mosaic typifies the daily output of the Advanced Vidicon Camera System obtained by the Tiros Operational Satellite (TOS) system for use by the U.S. Weather Bureau in preparing daily worldwide meteorological analyses and forecasts."

How high was the camera and how much of the Earth does each frame cover?

ESSA III

"The ESSA III satellite was in a 750-mile altitude, circular orbit, inclined approximately 79° retrograde to provide Sun synchronism. Each picture covers approximately 3 000 000 square miles of the Earth's surface. In order to provide geographic location of meteorological phenomena, each picture was routinely gridded at ESSA in a latitude-longitude matrix. The South Pole and the Antarctic region are clearly visible in the concentric circles in each of the lower pictures. Africa and the Near East are readily recognized in the central portion of the mosaic.

"The significance of this group of photographs lies in the fact that it represents a truly operational product of the space age and the result of more than 6 years of research and development work in the Tiros and Nimbus programs."

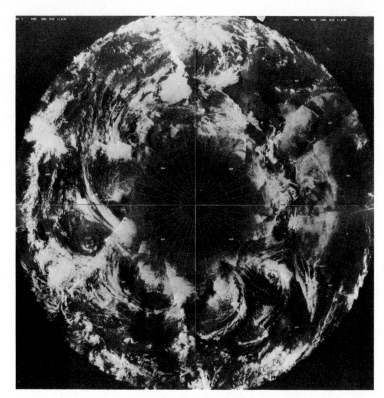

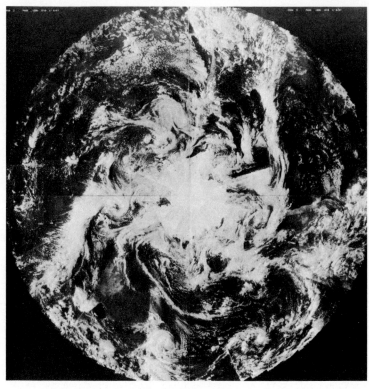

North Polar View

South Polar View

Fast Service for Airmen

"One hundred and fifty vidicon camera images collected by the ESSA III satellite were combined in the pictures on this page to reveal the cloud cover over the entire Earth on January 6, 1967," CHARLES L. BRISTOR, Chief, Data Processing and Analysis Division, at the National Environmental Satellite Center, tells us. "Each (800 scan line) image signal has been converted into 640 000 separate digital brightness samples. The fastest computer presently available is used to partially correct each image for variations in solar illumination and camera-lens vignet-

ting. The pictures are cropped to reduce overlap and then mapped in full resolution on northern and southern hemispheric polar stereographic projections (shown above). An overlapping Mercator mapping of regions bracketing the areas is below.

"Such routine mapping is creating montage images for timely use in weather briefing for pilots flying transoceanic routes and for use in improving weather analyses and forecasts. It also is providing vital input for further computer efforts toward the automatic extraction of information."

Arabia   India   North America   Africa

Australia   South America

Six Disturbances Seen Simultaneously

Applications Technology Satellites (ATS) are stationary with respect to the subsatellite point on the Earth because they rotate with the world. This and the pictures on the next four pages were obtained from such a satellite.

"The positions of six tropical disturbances were observed simultaneously on September 14, 1967, in this picture from the ATS–I satellite over the Pacific," says the Applications Technology Satellite Program Manager, JOSEPH R. BURKE.

"Beulah was developing in the Caribbean, Monica was west of Mexico, and Nanette was developing south of Baja California. At the same time, Sarah had started south of Hawaii and Vera and Opal were off the coast of Japan. As in many other ATS pictures, the intertropical conversion zone is visible along the line of the equator. One of the chief advantages of geostationary satellites is that they permit the meteorologists to view a very large area simultaneously."

A Day Passes

a Stationary

Photographer

7:05 a.m.

The ATS–I satellite transmitted these pictures of night and day December 11, 1966. The times given are Eastern Standard, and the photos show the changing cloud pattern over the visible hemisphere.

"These photos show the great advantage of viewing the Earth's storm clouds from a satellite in geostationary orbit," says VERNER E. SUOMI of the Space Science and Engineering Center at the University of Wisconsin. "Here the weather moves—not the satellite. The first photo shows the early morning sunlight over South America. A storm which looks like a white comma with its tail pointing west-southwest can be seen off the coast of Chile. By 3:45 p.m. and 5:45 p.m., the tail of the comma (really a cold front) has rotated a quarter turn to the north and the whole storm has moved farther east and south. Changes in the structure of another storm off the coasts of Oregon and Washington can be seen as the storm strikes the mountains. The snow-covered Sierra Nevada range can be seen northwest of Baja California, clearly visible in the 1:36 p.m. and 3:45 p.m. photos.

"Subtle changes in the convective clouds such as thunderstorms as the day proceeds are also visible over the Amazon and portions of the oceanic tropics. Even the equator appears to be visible due to the reduced cloudiness of the equatorial dry zone.

"Weather is air in motion. A camera on a satellite, stationary with respect to the Earth's surface, allows us to view the weather in a highly useful new dimension—time."

The camera that produced these pictures was at an altitude of about 22 300 miles, and was a "spin scan cloud camera." In reality, says ROBERT J. DARCEY, Applications Technology Satellite Project Manager at the Goddard Space Flight Center, NASA, "it is not a camera at all. It is a reflecting telescope with a pinhole aperture.

"Light which enters the pinhole is focused on a photomultiplier tube and its intensity is converted into electric impulses. The impulses are transmitted to the ground for processing on the terminal equipment which converts the impulses into photographic images. The east-west scan of the camera is accomplished by simply leaving the aperture open at all times, hence utilizing the satellite spin. The north-south scan is accomplished by mechanically tilting the camera after each line has been generated. Two thousand lines are required for one picture."

Recording each of the seven pictures here took 20 minutes.

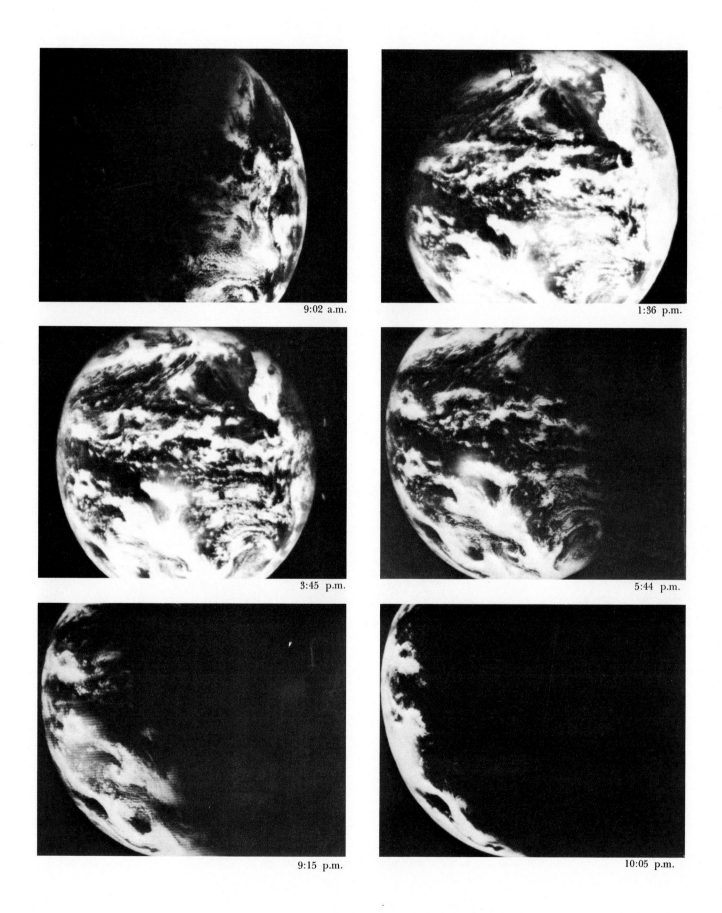

9:02 a.m.

1:36 p.m.

3:45 p.m.

5:44 p.m.

9:15 p.m.

10:05 p.m.

Storms Are Photographed
From Their Cradles Until They Die

January 2, 1967

January 3, 1967

January 4, 1967

January 5, 1967

January 6, 1967

January 7, 1967

"This sequence of pictures taken by ATS–I," MORRIS TEPPER, Deputy Director, Space Applications Program, NASA, reports, "gave meteorologists one of their first opportunities to study, in detail, the life cycle of a cyclonic storm as portrayed in cloud-cover pictures. The storm can be seen as it developed January 2, 1967, in the left part of the picture just above the Equator. On succeeding days, it is seen to progress in a northeasterly direction until it reaches a peak on January 4 and 5. The storm then may be seen in the upper center of the Earth's disk. The storm's clouds were very well developed, indicating it had reached its full maturity. On January 6 and 7, the storm began to dissipate, its cloud structure becoming more diffuse. The organized circulation can no longer be seen in the final picture.

"Many storms such as this have been observed and studied during the lifetime of ATS–I. The dynamic character of storm development becomes even more striking when these pictures are viewed as a movie."

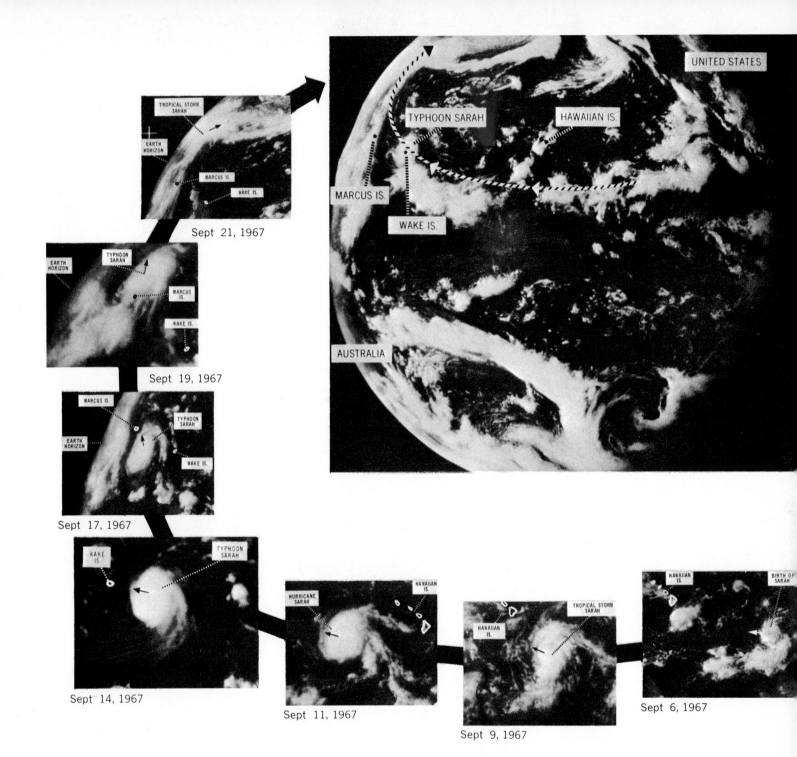

Sept 21, 1967

Sept 19, 1967

Sept 17, 1967

Sept 14, 1967

Sept 11, 1967

Sept 9, 1967

Sept 6, 1967

"Typhoon Sarah" (above), JOHN F. CLARK, Director, Goddard Space Flight Center, NASA, recalls, "caused considerable damage to Wake Island the afternoon of September 16, 1967. As can be seen in this sequence, and particularly in the pictures taken on September 14 and 17, the storm passed Wake Island during the period of its peak intensity. The severest part of the storm, the right forward quadrant, lashed Wake on September 16. The overall track of the storm, and specifically the September 14 picture, was used to provide reliable advanced warning to the island of Wake. ATS–I had the capability for continuously viewing the same portion of the Earth and its atmosphere during daylight from its vantage point over the Pacific. Sequences such as this are of considerable interest, not only to the tropical forecaster but also to research meteorologists who can use this kind of data to study the behavior of tropical storms and the conditions surrounding their birth and death."

TO THE MOON

AND BEYOND

TO THE MOON AND BEYOND

THE CAMERA SYSTEMS DE-
veloped for lunar and planetary work involved a
marriage of photographic and radiotelemetry techniques. By the time it was possible to put cameras
into space near the Moon and planets, pictures
could be transmitted by radio with very little de-
gradation from the transmission process. To date,
all pictures taken by unmanned spacecraft of the
Moon and Mars have been returned to Earth by
radio-propagation techniques.

Although man had studied the Moon for centuries, no one had ever seen its far side until this
decade. It was natural that one of the first things
that occurred to man for use of the camera in
space was to record the far side to compare it with
the visible face. From a scientific point of view,
the unknown nature of its crater-pocked surface had also caused considerable interest in the
small-scale features, and in determining the texture and composition of the surface.

The first Russian photographs of the Moon obtained with Luna III in October 1959 were crude
pictures of a portion of the far side and portrayed
the gross similarities and differences between the
far and near sides. The U.S. missions concentrated
on obtaining the first closeup look at the near
side; this was accomplished by Ranger VII in July
1964.

The Moon is particularly difficult to photograph
because it is a rather dark body with no atmosphere to diffuse light. Every object photographed
is therefore either in bright light or dark shadow,
resulting in extremely contrasty pictures. The
surface of the Moon also has the strange photometric property of reflecting light selectively in
the direction of the light source. Because of these
characteristics, pictures taken when the Sun is high
would show very little topographic detail; that is,
there would be very little contrast due to terrain

slope, and the pictures would be dominated by
shadow contrast. Pictures taken near lunar sunrise or sunset, however, emphasize slope contrast
by eliminating the backscattered light. Thus a
spacecraft sent to obtain high-resolution photographs must be aimed within a fairly narrow
range of lighting zones.

It was in December 1959 that the National
Aeronautics and Space Administration formulated
guidelines for a lunar program with objectives
that included high-resolution images of the lunar
surface. In a guideline letter to the Jet Propulsion Laboratory, inaugurating the effort later
known as the Ranger Program, it was stated:

"The lunar reconnaissance mission has been
selected with the major objective in mind being
the collection of data for use in an integrated
lunar-exploration program. Of the several specific experiments reviewed for assignment to the
early flights, the transmission of high-resolution
pictures of surface detail appears to be the most
desirable. It is, therefore, requested that your
program be directed to the consideration of a payload containing a picture transmission system
which will acquire and transmit a number of
images of the lunar surface. The system should
have an overall resolution of sufficient capability
for it to be possible to detect lunar details whose
characteristic dimension is as little as 10 feet."

Nine Ranger missions were flown; seven of the
spacecraft carried cameras. Rangers III, IV, and
V each carried a camera with a telescope of 40-inch
focal length and several other scientific instruments. Rangers VI through IX were designed
solely to obtain closeup photographs; each carried six television cameras of varying focal lengths
to provide both wide- and narrow-angle views of
the surface on approach. Because of develop-
mental problems, Rangers III through VI failed to

provide any photographic information. Rangers VII, VIII, and IX produced more than 17 000 high-quality photographs, giving man his first close view of the lunar surface. Included in these photographs were detailed images of three different portions of the Moon, two in widely separated mare areas and one inside the crater Alphonsus. Perhaps the last series of pictures was the most striking, for it showed a variety of features that included highland areas, a central peak in the crater, and the floor of this ancient crater. Ranger photographs were hailed by scientists for their clarity and detail, with the last few frames from each mission being as much as a thousand times sharper than any obtained from telescopes on Earth. A number of theories were confirmed, many were rejected, and new theories developed as a result. By far the most dominant topographical feature on the Moon was shown to be the crater, extending in diameter from hundreds of miles down to a few inches, in ever-increasing numbers toward the smaller scale. Past arguments about whether craters had been formed by impact or by volcanic activity were fused into combined theories indicating the probability of both types of formation. Perhaps of most significance was the fact that at high resolution, three widely separated areas on the Moon showed features of a very similar nature.

While the Ranger Program was getting underway, plans were laid for two other programs to extend observations of the Moon with cameras. The Surveyor Program called for soft landing of spacecraft able to observe minute detail on the surface, to view local topography as a man might see it, and to make other scientific measurements. The Lunar Orbiter Program called for putting a camera in orbit about the Moon to observe areas of the surface at extremely high resolution and to survey the entire lunar surface.

On the very first mission, Surveyor I landed successfully and returned some 10 000 pictures before sunset on its first lunar day. Included in these pictures were closeups of the spacecraft itself, the soil that had been disturbed by the spacecraft, the features on the distant horizon, rocks and craters nearby, some of the planets and stars, and the setting Sun as it slowly dropped over the horizon. The camera continued to operate into the lunar night and took some pictures with the light of earthshine. During the second lunar day, it returned 899 additional pictures.

Surveyor I was joined by Surveyor III on the Moon after Surveyor II failed to operate properly and crashed onto the surface. Surveyor IV also experienced difficulty, but Surveyors V, VI, and VII succeeded in every respect. Surveyor III was significant because, in addition to the camera, it carried a surface sampler which allowed scientists to manipulate the soil under the view of the camera. In a way, this spacecraft represented the placing of man's hands as well as his eyes on the surface of the Moon, for scientists were able to command the sampler on the basis of what they saw, much as a scientist operates in his laboratory.

Surveyor V will be remembered for its first measurements of the constituents on the lunar surface, although it also returned 19 000 photographs. An alpha-backscattering instrument was able to record the elemental composition of the Moon's surface at one location to an accuracy of a few percent, indicating that it is much like many basalts here on Earth. Again the camera served to provide related information on the surface and the placement of the instrument.

Surveyor VI continued the exploration of the lunar surface with the successful landing near the middle of the Moon as seen from Earth. It repeated the survey of the surrounding terrain and the topography close at hand, transmitting a record 30 065 pictures during the first lunar day. It also gave a second set of readings on the composition of the lunar surface, indicating that the material was essentially the same as that in the region farther east measured by Surveyor V.

A series of five Lunar Orbiter spacecraft have successfully photographed, in resolution of about 3 feet, 25 sites that appear to be suitable for manned landings, and from these, 5 potential sites for manned landings have now been selected.

In addition to the landing areas, many sites of high scientific interest have been photographed to a resolution of from 3 to 15 feet. The entire front face of the Moon has been covered at a resolution between 175 and 400 feet, providing topographic information that will allow very detailed geologic studies.

Some of the most exciting photographs obtained by Lunar Orbiter were oblique views of the surface. One of the earliest photographs showing the limb of the Moon also revealed the Earth in the distance, clearly showing both the terminator and the cloud coverage over the sunlit portion of the globe. Because there is no atmosphere on the Moon, oblique photographs with the telephoto lens are perfectly clear and distinct, even though features photographed were up to 75 miles away. The views of the steep crater walls and landscapes are very helpful to the scientific interpretations of topography when used in conjunction with photography obtained from a vertical view.

The series of automated spacecraft containing cameras that have been sent to the vicinity or surface of the Moon have provided detailed knowledge of the surface topography. When combined with other data, photographs have enabled us to examine the relationships between the Moon's topography, its specific mechanical characteristics, and its chemical composition. Photographs allow improved accuracy of mapping for the near side, and for the first time have provided coverage of the far side. Manned landing sites have been selected from photos and from measurements returned by unmanned spacecraft, and are being thoroughly studied both to insure the safety of the astronauts and to achieve the greatest scientific returns from man's visit.

Although much has been done, there are still many potential uses for cameras in exploring the Moon. Because shadowing is necessary to provide topographical detail, many areas of the Moon have not been photographed with high resolution. Under different lighting conditions, new knowledge will be obtained. When man sets foot on the Moon, it is certain that his cameras will return to Earth a wealth of new data.

Because of their greater distances from Earth and because of their atmospheric properties, we know far less about other planets in the solar system from observations by telescope than we know about the Moon. Only one spacecraft has successfully photographed a planet from nearby. This was Mariner IV, which flew by Mars in 1965 and returned 22 pictures. One of these showed an oblique view of the limb of the planet. Perhaps the most striking of the Mariner pictures was one showing many craters and a surface surprisingly like that of the Moon. Elevation changes of thousands of feet were noted and portions of the terrain seemed quite rough.

The future for planetary photography is tremendous. From telescopes on Earth, we know that many of the planets have strikingly different characteristics. The day will surely come when cameras will be flown to the vicinity of these interesting planets, and we will be able to view the results here on Earth. The next planetary mission planned by the United States with emphasis on photography is a pair of missions to Mars in 1969, which will not only provide photographic coverage of most of the surface at resolutions better than we have from Earth, but will also provide closeup looks that will greatly expand our knowledge of Mars.—ORAN W. NICKS

The Moon That Man
Had Never Seen Before

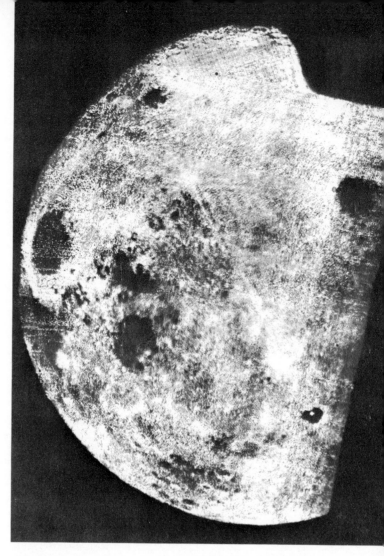

After centuries of wondering what the other side of the Moon looked like, man finally got his first, engrossing glimpses of it in the fall of 1959.

The first prints of those historic Soviet photos taken by Luna III were not quite as clear as the two shown here. These have enhanced quality, explains WILMOT N. HESS, Director of Science and Applications, Manned Spacecraft Center, NASA, because "they are a combination of various negatives that were screened, combined, and rectified by workers at the University of Arizona's Lunar and Planetary Laboratory.

"Luna III, launched October 4, 1959, was the first space probe to photograph another planetary body and transmit the pictures back to Earth," Hess continues. "The photographs taken covered more than 10 million square kilometers on the far side of the Moon. This meant that about 80 percent of the lunar surface had been covered by photography of various resolutions.

"Features identifiable on the photos here include Tsiolkovsky [dark crater with bright peak at lower right], Mare Moscoviense [dark area at upper right], Mare Australe [large, irregular dark area on lower edge], Mare Fecunditatis with Mare Crisium above it [left edge], Mare Smythii [large dark area with two bright peaks], Mare Marginis [above Mare Smythii], and Mare Humboldtianum [dark area at top left]. Tsiolkovsky and Mare Moscoviense were viewed for the first time on Luna III photography. The Soviets

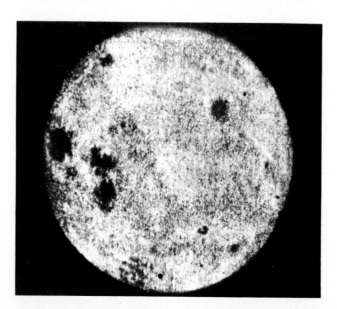

also interpreted the bright area above Tsiolkovsky as a mountain range. (The higher resolution Lunar Orbiter photography has since shown this area to be heavily cratered, but not mountainous.) Luna III's photographs indicated that the Moon's far side has no large mare areas and contains much more continental area than the side facing Earth."

"The Soviet space probe Zond III," one of whose pictures is shown on the facing page, says Hess, "was launched on a mission that photographed the 20 percent of the lunar surface that had not previously been viewed. Picture taking began at 1:24 universal time on July 20, 1965, at an altitude of 11 570 kilometers. During the following 68 minutes, 25 photos were taken, and the altitude gradually decreased to 9220 kilometers, then increased to 9960 kilometers. The area photographed extended from the western edge of the visible side of the Moon across the far side to the terminator.

"The large photograph shows the equatorial and northern portions of the lunar far side. The large dark area on the right is Mare Orientale, which is

located on the western limb of the Moon when viewed from Earth. (The orientation of cardinal directions is in accordance with the Astronautical Convention adopted by the IAU General Assembly of 1961.) The dark area at the right edge of the picture is the crater Grimaldi, which is on the visible side of the Moon.

"With this Zond III photography and that of the earlier Luna III, only an area at the south polar region of the far side of the Moon remained to be viewed.

"The photography taken by the Zond III camera was of a higher quality than that of Luna III's camera, and confirmed the initial impressions that the far side of the Moon has fewer large maria and much more continental area than the visible side."

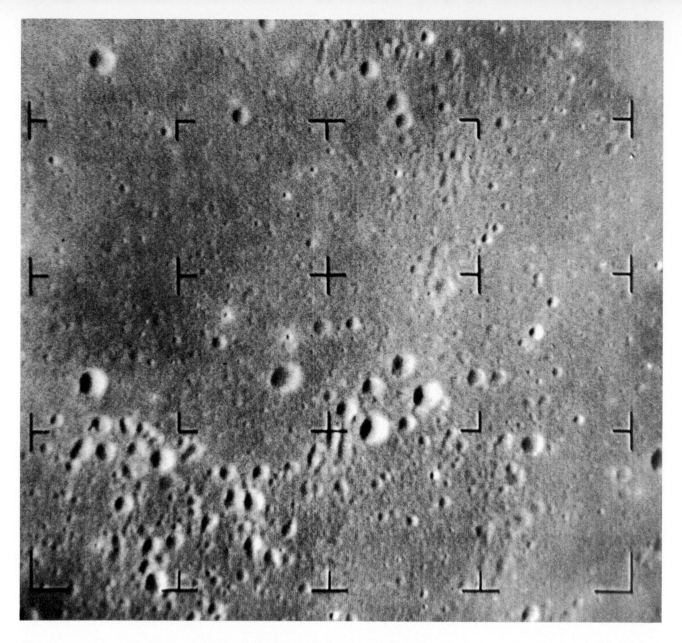

Rangers Bring the Moon Up Close

Astronomers were jubilant at the clarity of the thousands of closeup photos of the Moon taken in 1964–1965 by the TV cameras on Rangers VII, VIII, and IX during the final moments before impact.

Of the four photos shown on these pages and on the two following pages, the one above, taken on July 31, 1964, by Ranger VII, was "the first high-resolution look at the surface details within faint crater rays," said HARRIS M. SCHURMEIER, former Ranger Project Manager, Jet Propulsion Laboratory. "It shows an area about 12 miles wide, which the International Astronomical Union renamed 'Mare Cognitum'—'Known Sea'—as a result of this mission. Ranger VII confirmed that there were lunar areas

topographically acceptable as manned landing sites."

Of the same picture, GERARD P. KUIPER, Director, Lunar and Planetary Laboratory, University of Arizona, noted: "The photograph shows several clusters of secondary impact craters, found to be bright at full Moon, which are accompanied by short, diffuse ray elements. Some of these clusters belong to Tycho's ray system [prominent center group] and some to Copernicus' [at top, lower left, and lower right]. Evidence indicates that Copernicus and Tycho were both caused by impacts of comets entering the planetary system on parabolic orbits, the secondaries seen on the picture having been caused by associated cometary debris. This photo, in addition, shows numer-

ous small, shallow collapse depressions in the mare floor. Some are nearly linear along well-known structural directions.

"Ranger VII's records," Kuiper concluded, "were the first to close the gap between Earth-based photography (best resolution, 0.4 kilometer) and fine-structure data derived indirectly from thermal, radio, and radar observations. The overlapping records proved the consistency and reliability of the data."

The Ranger VIII photograph above shows the sister craters Ritter and Sabine, on the edge of Mare Tranquillitatis. RAYMOND L. HEACOCK, Chief, Lunar and Planetary Instruments Section, Jet Propulsion

Laboratory, recalled that "it was one of 7137 photographs that Ranger VIII returned before impacting within 30 kilometers of its target in Mare Tranquillitatis, on February 20, 1965. Taken from an altitude of 243.4 kilometers, it covers approximately 25 square kilometers.

"The resolution," says Heacock, "is about 10 times the best Earth-based resolution, thus revealing considerably greater detail than ever before about the structure of the Hypatia Rilles and the flat-bottomed craters Ritter and Sabine. While the rilles appear to be similar to the graben found on Earth, the cause of the faulting is not revealed. Several north-northwest gouge features are the result of secondary

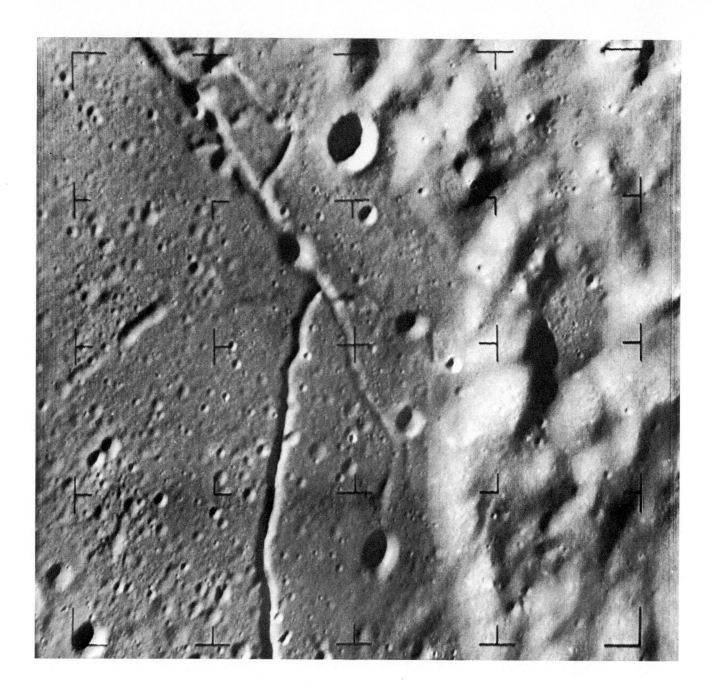

impacts of ejecta derived from the crater Theophilus.

"Ranger VIII impacted within the Apollo landing zone and achieved a terminal resolution of approximately 1.5 meters."

Harold C. Urey, Nobel laureate and professor at the University of California, San Diego, said of the Ranger IX photo above: "This picture shows a portion of the floor and crater wall of Alphonsus. The floor is covered with many craters of various sizes, some sharp and hence new, others less distinct and partly filled with fragmented material. The walls have fewer craters, and this probably means that slumping of the wall has filled them. Crevasses are evident, and evidence for slumping exists. The larger crater near the top is undoubtedly collisional in origin. Three craters are surrounded by dark halos and were produced by eruptions from the lunar interior. Exceptionally bright, sharp peaks can be seen on certain mountain tops. Are they possibly metallic objects that resist particle erosion? The crater floor and the smooth area in the walls at the right are usually assumed to be lava. Are they possibly the beds of temporary lakes?"

Ranger IX pictures were seen on TV in the U.S. and Europe as received "live from the Moon."

Concerning Ranger IX's last photograph [shown above], WILLIAM H. PICKERING, Director, Jet Propulsion Laboratory, commented that "it was taken one-quarter second before impact, at ⅓-mile altitude, and shows an area 200 feet from top to bottom and 240 feet from left to right on the floor of the crater Alphonsus.

"The surface," Pickering said, "is dominated with impact craters with very few positive relief features.

The low Sun-elevation angle (10.4°), combined with the surface resolution, would have permitted positive detection of surface blocks on the order of 0.2 to 0.3 meter in diameter.

"Ranger IX, which yielded 5814 Moon-surface pictures, provided confirmation of the dominant role that impact cratering has had in establishing the small-scale topography in both mare and large highland craters. Luna IX and Surveyor I provided additional confirmation of this process.

"The television camera that took this picture had a focal length of 76 mm, a field of view of 2.1°."

First Soft Landings on the Moon

"A chill and friendless thing," A. C. Benson called the Moon, and history's first on-the-spot photographs of its bleak surface clearly bore him out.

Early in 1966, a Russian space probe, Luna IX, made the initial soft landing there. One of the pictures that its TV camera took and transmitted to Earth is shown above.

Luna IX landed approximately 100 kilometers northeast of the center of the crater Calaverius on February 3, 1966. The next day, it transmitted its first lunar panorama, part of which appears at left. Of this picture, JACK GREEN, of Douglas Advanced Research Laboratory, said:

"The photograph, with a resolution of some 1.5 to 2 mm at a distance of 1.5 meters, shows: (1) no thick dust; (2) both rounded and angular rock fragments, either concentrated on crater rims or randomly distributed; (3) small ridges and linear open fractures; (4) numerous small craters, some with inner slope angles of over 40°; and (5) a generally cellular-to-granular surface texture.

"Not only did Luna IX prove the lunar surface strong enough to support astronauts but it also showed good evidence of pitting on the sides and surfaces of rocks, suggestive of vesiculation. Luna IX photography weakened exotic cotton candy or whisker theories advocating a tenuous lunar surface, and, in my opinion, strengthened more conventional theories of volcanological processes being responsible for shaping the major surface features of the Moon."

A few months later, in June 1966, the U.S. Surveyor I soft-landed on the Moon. It used many new elements, including throttleable vernier rockets, sensitive velocity- and altitude-sensing radars, and an automatic, closed-loop landing system. During the following 6 weeks it transmitted 11 150 high-resolution pictures back to Earth.

Surveyor I's camera system had a variable iris, changeable filters, and a rotating mirror assembly, which allowed the camera to look in almost any direction and take pictures under various lighting conditions, in either black and white or in color. Video pictures with 200-line resolution and with 600-line resolution were possible; the first with a quick-look mode and the capability of transmission with a low-gain antenna; the second for use with the directional antenna and the high data rate.

The bright lunar vista stretching across the bottom of the facing page is part of a spherical mosaic prepared from more than 200 pictures that Surveyor I's camera took on June 13, 1966. One of the spacecraft's three feet protrudes in the foreground.

EUGENE M. SHOEMAKER, Chief, Astrogeology Branch, U.S. Geological Survey, commented on the scene:

"It shows the intricately cratered surface of the Oceanus Procellarum. It is a gently undulating surface pockmarked with craters, ranging from a few centimeters to several hundred meters in diameter, and littered with blocks and fragments, ranging from less than a millimeter to more than a meter across. The craters and the fragmental debris were probably formed by bombardment of the lunar surface by meteoroids and by pieces of the Moon itself, hurled through space from larger craters.

"The pitted appearance of the lunar surface [which one observer has likened to that of a World War I battlefield] is enhanced by the presence of long shadows, extending away from the spacecraft's foot and the more prominent blocks, due to the low angle of incidence of the rays of the evening Sun."

Surveyor I's TV camera had a zoom lens with focal lengths ranging from 25 to 100 millimeters. One of its most effective near-view photos is reproduced at right above.

STEPHEN E. DWORNIK, Surveyor Program Scientist, NASA, reported: "This photograph shows a 12- by 18-inch block located about 13 feet from the camera. It is subangular in shape, with facets slightly rounded

at the edges and corners. The lighter colored part of the rock is the more resistant to erosion, and therefore distinctive. Granularity of the block is not evident, but it shows mottling. Intersecting fracture planes resemble cleavage planes produced during static flow of rock under high shock pressure. The edges of the fractures also exhibit a rounding. No fracturing or crevicing is evident in the lighter-colored part of the block.

"The block lies near many smaller blocks," Dwornik added, "and it is likely that all the blocks and fragments in the photograph are of the same origin. Both the smaller fragments and the block itself are partly submerged by the younger, finer material composing the lunar surface. The grain size of the lunar surface material cannot be determined because it is below the resolution of the camera."

Footprints in the Lunar Soil

On June 2, 1966, some 63½ hours after being launched from Cape Kennedy, Surveyor I sat its 596-pound weight down on the Moon without apparent difficulty. It had landed on a dark, relatively smooth, bare surface north of the crater Flamsteed, in Oceanus Procellarum. The geographic coordinates of the site, encircled by hills and low mountains, were 2.41° S, 43.34° W.

In reference to the mosaic view above, A. R. LUEDECKE, then Deputy Director, Jet Propulsion Laboratory, wrote, "The ridge of surface material pushed up during landing by the outward motion of the footpad, 12 inches in diameter, is well outlined by shadow, for the Sun was approximately 10° above the horizon when the pictures in the mosaic were taken. The contrast in texture and albedo between the undisturbed surface and that disturbed by the footpad is noticeable, particularly above the footpad. Both of these features indicate that the lunar surface material at the Surveyor I site is not hard rock but a structure disturbed appreciably when subjected to

the rather low loads of the order of 5 pounds per square inch.

"This photograph is a mosaic of separate frames [narrow-angle, 6° field of view] taken by the Surveyor television system. Most of the frames were digitized, processed on a computer to correct for system-frequency response, and then converted back to image form." It was a technique used throughout that mission and the Surveyor missions that followed.

"The lunar material seen in the picture on the opposite page is primarily material thrown out from beneath a Surveyor I footpad during landing," said EDWARD N. SHIPLEY, of Bellcomm, Inc., "although some of the original surface layer, which is lighter in color, is visible. The darker color of the subsurface material was unexpected, and there is currently no widely accepted explanation for it.

"Geometrical analysis," Shipley continued, "shows that the bottom of the footpad's depression is between 1 and 2 inches below the undisturbed lunar

surface. Additional data from the spacecraft teleme-try give the magnitude of the forces exerted on the surface during landing. The size of the depression, compared with the magnitude of the impact forces, provides a quantitative measurement of the strength of the lunar surface material. From this informa-tion, further analysis has shown that there is no danger that either an Apollo spacecraft landing on the Moon or an astronaut walking on it would sink hazardously into the surface."

"This remarkable self-portrait [above] of Surveyor I is one of 144 pictures taken during its first 24 hours of operation on the Moon," said W. E. Giberson, formerly Surveyor Project Manager, Jet Propulsion Laboratory. It was one of the earliest of the 600-line high-resolution television pictures that the space-craft took. Initial photographs were made in the system's 200-line scanning mode.

"At upper left," continued Giberson, "is footpad 3 (12 inches in diameter), attached to the landing gear and shock absorber. Below the footpad are one of the spacecraft's omnidirectional antenna booms,

54

its helium pressurization tank (14 inches in diameter), and the tank's valves, regulator, and lines.

"Details as small as a fiftieth of an inch can be seen. Most of the lunar surface particles are even smaller."

Surveyor I landed on the Moon at a vertical velocity of approximately 10 feet per second, rebounded about 2½ inches, then came to rest. Its footpads, in the process, slightly disturbed the surface, as shown in the picture at top left.

"This photograph was taken on June 13, 1966, at 19:45:35 GMT," wrote HOWARD H. HAGLUND, Surveyor Project Manager, Jet Propulsion Laboratory. "The Sun at the time was 10° above the western horizon. The view is of the outboard edge of footpad 2 of Surveyor I, and of the lunar surface material beyond it.

"The arc of lighter-colored material nearest the footpad is part of the ridge pushed up by the pad. It shows lumps, or 'clods,' up to an inch in diameter, produced by the pad pressure during landing. These are apparently made up of fine particles (less than one-fiftieth inch in diameter).

"The photograph is scientifically interesting in that it shows that the lunar-surface material is co-

hesive. Its mechanical properties are roughly similar to those of damp garden soil."

Describing the picture at top right, THOMAS GOLD, Director, Cornell University's Center for Radiophysics and Space Research, said: "The sharp cut in the lunar surface material directly underneath Surveyor I was caused during the landing by a thin heat shield, which cut into the soil as a cookie cutter would. When the springing of Surveyor's legs restored the spacecraft to its normal height, the 'cookie cutter' was withdrawn.

"The picture demonstrates," said Gold, "that the material is fine grained and cohesive to the extent that it can be cut and leave a vertical face. Clumps thrown out at impact are clearly visible, so it is a crumbly, crunchy surface.

"A dry material in vacuum has such properties only if it is composed of particles mostly smaller than 0.01 millimeter. A coarser aggregate, like sand, would slide and leave only incline slopes, never vertical ones.

"The picture strengthens the case that the porous overlay known to cover most of the Moon consists of cohesive, fine rock dust, and not either sand or solid, bubbly lava."

Surveyor I Takes

a Look Around

Surveyor I's camera had a lens of variable focal length and could be pointed by radio command from Earth. This allowed scientists to choose their subject and the most suitable light and lens setting for photographing it. Surveyor's scanning of its horizon, affording man his first look around the land-scape of another world, provided the views on this page and the three following pages.

"A surface born in violence!" exclaimed RALPH B. BALDWIN, a noted astronomer, referring to the lunar scene above. "Prominently displayed on this Surveyor view of a mare is a small crater, perhaps 3 meters across, dug by the impact of a small meteorite. Scattered essentially at random over the surface are numerous rocks ejected from other, larger craters.

"In the distance," Baldwin pointed out, "is a strange wall of rocks. Study shows that these rocks lie, in part, on the near wall of a crater some hundreds of meters in diameter and, in part, in a sort

of raylike extension of the crater wall. They were thrown out when the crater was formed. It is not known whether they represent true subsurface, stony materials, or are pulverized materials consolidated by the shock waves from the violent impact."

A telephoto view of the rocks referred to is shown above. BENJAMIN MILWITZKY, Surveyor Program Manager, NASA, said of it, "This remarkable photo shows a field of large rocks several hundred feet from the spacecraft. They range from about 3 to 6 feet in diameter, and appear to have been excavated from beneath the lunar surface and hurled outward by the impact of a large meteorite.

"Ejecta of this type are brighter than the undisturbed mare surface, and may contribute to the 'rays,' radiating from large impact craters, seen through Earth-based telescopes.

"The fact that the rocks came to rest on the surface," Milwitzky continued, "suggests that the surface, at least in this area, must have appreciable bearing strength, adequate to support manned spacecraft. This evidence is consistent with data from Surveyor's landing-gear strain gages and pictures of its footprints. The large rocks, on the other hand, represent a significant hazard, which the astronauts will have to avoid by carefully maneuvering their spacecraft during its descent."

SURVEYOR I

A Far Mountain, a Near Crater

One of Surveyor I's most interesting telephoto views [at top] was of a mountain range about 16 miles northeast of the spacecraft. E. C. MORRIS, of the U.S. Geological Survey, described it as follows:

"The landing site was within a large, ancient crater, more than 60 miles in diameter, buried by mare material. "The mountain range shown here is part of the rim of this nearly buried crater. The highest peak rises more than 1300 feet above the lunar surface, but only the upper 600 feet project above the near horizon, which is little more than a mile away. The observable crest of the range extends approximately 3 miles along the horizon."

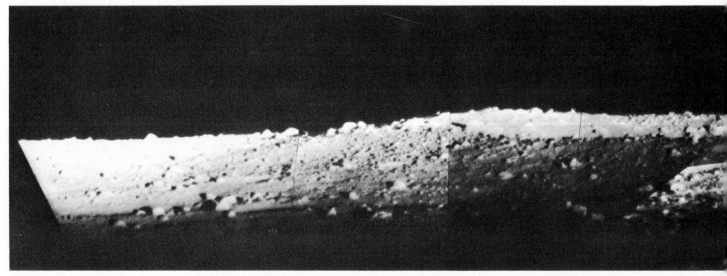

The mosaic of TV pictures below forms a broad vista of the rim of a low crater several hundred yards distant from the landing site of Surveyor I.

ZDENEK KOPAL, of the University of Manchester, in England, made the following points about it: "What makes this photograph unique is the vantage point from which it was taken. The crater, one of hundreds of thousands of this type on the Moon, is much akin to the well-known Meteor Crater in Arizona; namely, caused by the impact of an outside mass.

"However, from the shape of the ramparts (and boulders dispersed around them), we may conjecture that the crater itself is likely to be of secondary, rather than primary, origin. Also, whereas the Meteor Crater is probably not more than 100 000 years old, this lunar one may well be 10 times older. So slow are the grinding processes of cosmic obliteration on the Moon!"

The Long Lunar Night Descends

The pictures on these pages were taken toward the close of two different lunar days in 1966, by Surveyor I (above), in mid-June, and by Russia's Luna XIII (opposite), the day after Christmas.

"As the Sun approaches the lunar horizon," GEORGE H. SUTTON, University of Hawaii, said of the photos above, "lengthening shadows emphasize the pockmarked and rubble-strewn character of the surface. Raised rims of several craters and angular outlines of many surface fragments stand out sharply. In addition to giving a clear impression of the nature of the lunar surface, the long shadows in such pictures, taken at low Sun angles, can be measured to determine shapes and sizes of surface irregularities, obtaining accurate estimates of the size distribution of fragments and the shapes of craters."

The photographs have additional value. "Comparison of the craters in them with craters obtained in Earth-bound experiments of high-speed impacts in simulated lunar materials," Sutton continued,

"permits estimation of the mechanical properties of the Moon's surface."

The pair of Luna XIII photographs shown on page 61 are the last two pictures of a 360° panoramic view of the Moon's surface.

"The horizon appears to slope because of the tilt of the television camera," explained JOHN A. O'KEEFE, Assistant Chief, Laboratory for Theoretical Studies, Goddard Space Flight Center, NASA. "In the lower left and center of the top picture is a beam that was extended from the spacecraft to implant a penetrometer, measuring soil resistance to penetration. At left center in the bottom picture is the rounded end of one of the 'petals' that covered Luna XIII's instruments through the landing, and then unfolded.

"Note the change in the texture of the landscape about 3° below the horizon," O'Keefe added. "This is stated to be the rim of a crater in which Luna XIII rests. And note the blocks beyond, toward the horizon."

Self-Portraits of a Star Performer

"Alone on a desolate plain of the Moon's Sea of Storms (Oceanus Procellarum), Surveyor I stands quietly, its job well done," wrote HOMER E. NEWELL, Associate Administrator, NASA, in describing the photograph above. "Here is a picture of its own making. Surveyor casts a lengthening shadow as the long lunar day nears its end.

"The mosaic of approximately 52 photographs shows the rough texture of the Moon's cratered and pitted surface, on which rocks and boulders may be seen scattered here and there. Surface temperatures, which at lunar noon had risen to 235° F, are now falling slowly, a mere hint of the approaching plunge to 250° below zero after sunset."

[One should not be misled by the sinusoidal appearance of the lunar horizon in the mosaic. This occurred because Surveyor I's camera was tilted, in order to observe the spacecraft's feet better. Thus, when the camera was pointed in the direction of the tilt, the horizon appeared higher than it was. When the camera was pointed in the opposite direction, it looked slightly upward and the horizon appeared lower. In reality, the vicinity was largely flat.]

Surveyor I survived the 14 earthly days of its first lunar night, took and transmitted 10 388 pictures during its first lunar day, 812 the second, and sent its last photograph on July 13, 1966, as the Sun was setting on the Moon.

"Because of battery failure," Newell continued, "presumably caused by the bitter cold of the lunar night, Surveyor I can no longer send earthward pictures of its lonely vigil. However, with solar power,

its radio continued to function during each lunar day for 8 months after landing, answering simple questions about the status of the spacecraft.

"Surveyor I stands physically on the Moon, an enduring monument to its creators, a solitary artifact of men who live on another body of the solar system, a quarter of a million miles away, but its true resting place is in the pages of history, where even now is being inscribed man's conquest of space."

"The performance of Surveyor I met or exceeded expectations in all areas," commented ROBERT J. PARKS, Assistant Laboratory Director for Lunar and Planetary Projects, Jet Propulsion Laboratory. "After its successful soft landing, a planned series of 200-line pictures was transmitted, using an omni-

antenna. The high-gain antenna positioning equipment was then exercised for the first time. It responded immediately, and was properly positioned. The switchover to the 600-line mode was then accomplished. It was in this mode that the many high-quality photographs were obtained.

"In the 600-line mode, during the first lunar day, nearly complete coverage of the surrounding area was obtained under a wide variety of Sun-angle and shadow conditions.

"Toward the end of the first lunar day," Parks concluded, "as the shadows lengthened, the Surveyor camera obtained the fascinating silhouette self-portrait at right above. This picture symbolizes in a very dramatic way the highly successful performance of all elements of the mission."

After the Sun
Went Down

"The solar corona in the photograph at top right was observed by Surveyor I, 16 minutes after sunset on the Moon on June 14, 1966," remarked GORDON NEWKIRK, of the High Altitude Observatory. "A bright coronal streamer is visible as a thin pencil of light extending out of the brighter inner corona, against which the lunar horizon is silhouetted. Halation in the optical system produces an apparent depression of the horizon immediately below the brightest portion of the corona. The faint disk in the upper-right corner and the bar extending to the lower-right corner are the omnidirectional antenna and its support structure, visible in the light of earthshine.

"This photograph demonstrated the feasibility of making observations of the solar corona from lunar-based observatories. With little or no modification, the Surveyor cameras can investigate the outer corona and inner zodiacal light, which are inaccessible to ground-based observatories.

"The fact that the corona is visible after sunset demonstrates that little or no residual atmosphere exists on the Moon. Detailed photometric comparison of these Surveyor observations with those made on the ground the same day will allow an exact upper limit of such a residual lunar atmosphere to be established."

In reference to the photograph at bottom right, LEONARD D. JAFFE, Surveyor Project Scientist, Jet Propulsion Laboratory, said, "This photograph was made during the lunar night, using light provided by the Earth. It shows a footpad and adjacent parts of Surveyor I against the darker background of the lunar surface. The indentation made by the footpad in the surface is also visible.

"As evidenced by this picture, it is possible to get useful photographs on the Moon at night, using earthshine for illumination. The surface is bright enough for human vision.

"The bright spot at the left," added Jaffe, "is a reflection of the Earth in the optical system. The small circular dish with radial stripes is a calibrated photometric target mounted on the spacecraft leg.

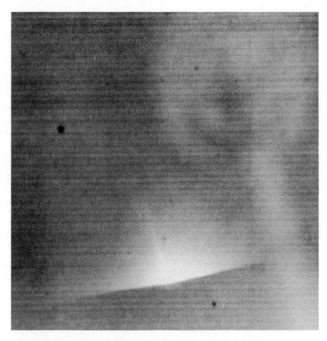

It permits accurate determination of the light intensity and surface brightness.

"This photograph was taken by Surveyor I's television system on June 14, 1966, at 16:35 Greenwich mean time. The iris aperture (which provided nominal focal ratios ranging from $f/4$ to $f/22$) was $f/4$. The exposure time (150 milliseconds in the normal mode) was 4 minutes."

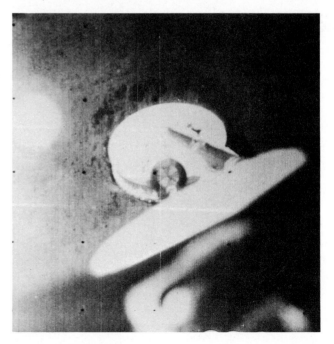

Hop, Skip, and Jump

Surveyor III bounced three times on landing, April 20, 1967, and slid down one steep wall of a crater. The multiple landings arose from a failure of its vernier engines to shut off as programed. The mosaic above shows the surface disturbances made by its three legs (1, 2, 3) and one of the thrusting vernier rocket engines (V) at the end of the first bounce. Footpad 2 hit the surface first, followed by 1 and 3. All three rebounded, but 2 hit again (2') while 1 and 3 were still in the air.

"These surface disturbances," explained SIDNEY A. BATTERSON, Dynamic Loads Division, Langley Research Center, NASA, "are located approximately 14 meters upslope and in an easterly direction from the final resting place of the spacecraft." Impact marks of similar shape and depth were observed where Surveyor III finally came to a halt and where Surveyor I landed.

"Since the landing sites are hundreds of kilometers apart," Batterson concluded, "and since Surveyor I landed on a relatively flat plane, whereas Surveyor III landed on the wall of a crater, the similarity in the behavior of the surface material at both sites suggests that the material is quite homogeneous over very large areas of the lunar surface."

Putting Man's Hands
on the Moon

Surveyor III's television camera could not view the lunar surface outside the crater within which the spacecraft had landed, but the sloping walls of the crater allowed it to view nearby features more clearly than would have been possible on flat terrain. The spacecraft took 6315 pictures between April 20 and May 3, 1967.

This Surveyor, unlike its predecessors, carried a remotely controlled surface sampler, a device for digging and otherwise manipulating the surface material in the view of the television camera. The sampler made 8 bearing tests and 14 impact tests on the lunar surface, dug 4 trenches, and picked up 3 objects. One of these, a small rock [see opposite page, top right], was gripped by the sampler's scoop with a pressure of at least 100 pounds per square inch without apparently crushing or breaking it.

"Through the camera eye of Surveyor III," commented S. C. SHALLON, Chief Surveyor Program Scientist, Hughes Aircraft Co., "we see at top right a lunar surface capable of supporting both men and their spacecraft in future exploration of the Moon.

"This photograph was transmitted from the Moon by Surveyor III on April 26, 1967. The circular impression was made by one of the three footpads on the last bounce of a three-bounce landing. The surface impression at bottom left was made by the 'scooper,' a digging device shown in an extended position here. These surface impressions appear similar to those that might be made in damp, fine-grained soil on Earth.

"We know, of course, that the lunar surface cannot be damp, because of the near-vacuum conditions there. But the results of the scooper's experiments, as viewed through Surveyor III's camera, together with other measurements made by the spacecraft, indicate that the Moon's surface does indeed have a consistency similar to that of damp, fine-grained terrestrial soil."

"The dark, longitudinal area seen in the photograph immediately below," explained MAURICE C. CLARY, Lunar and Planetary Instruments Section, Jet Propulsion Laboratory, "is a furrow in the lunar surface that resulted from a single trenching opera-

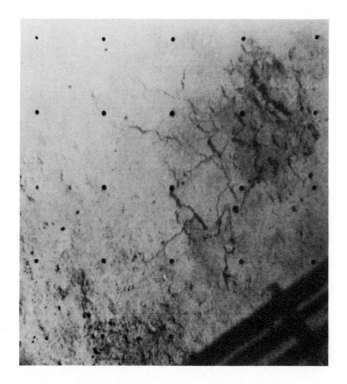 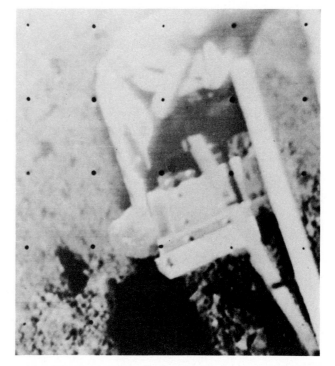

tion by Surveyor III's soil-mechanics surface sampler. Some of the lunar fragments along the length of the trench resemble terrestrial rocks. Further manipulation of the surface sampler, which included closing its scoop door on one of these objects, proved that most of the fragments were aggregates of more finely divided particles, similar to clods. That was not true of all, however, as the photograph at top right on this page reveals.

"In the photograph," Clary continued, "the majority of the lunar-surface material displaced during formation of the trench appears at the left. This is due to the slope of Surveyor III's landing site."

"The area [1 by 2 inches] where Surveyor III's surface-sampler scoop contacted the Moon's soil was slightly depressed, and some of the adjacent soil was cracked and slightly lifted," said ELMER M. CHRISTEN-SEN, of Jet Propulsion Laboratory, in regard to the photo at top left. "This and other Surveyor tests—surface-sampler operations, landing imprints, landing loads, vernier-engine firings, attitude-control jet firings, and deposition of small amounts of soil on the spacecraft—have revealed the lunar soil to be fine grained, granular, and slightly cohesive, with greater strength and density 6 inches below the top surface.

"At the Surveyor landing sites," Christensen declared, "the soil is amazingly uniform, much more so than at terrestrial areas so widely separated. Because of programs like Surveyor, man will explore the Moon, walking and driving his machines, with significantly fewer restrictions than on Earth."

"During bearing-test operations on May 1, 1967," recounted F. I. ROBERSON, Lunar and Planetary Sciences, Jet Propulsion Laboratory, "an object that appeared brighter than the surrounding surface was observed. A careful scraping action by the surface sampler dislodged the partially buried fragment. By alternately taking television frames and commanding the sampler, we brought the scoop slowly into position and closed the scoop door on the object. The result, shown at top right, was this photo of a lunar rock being picked up in the edge of the scoop's jaw.

"The soil-mechanics surface-sampler experiment, including picking up this rock, marked the first time in history that man has manipulated by remote control the surface of a celestial body other than Earth.

"This rock test also provided the first conclusive evidence that objects observed on the Moon through Surveyor's TV camera that looked like rocks were indeed rocks, and not clumps of dirt."

Relating Surveyor III to Its Surroundings

"The value of the data transmitted by the various successfully landed Surveyor spacecraft is considerably augmented," declares EWEN A. WHITAKER, Lunar and Planetary Laboratory, University of Arizona, "when actual locations of these craft can be found on Orbiter photographs of their respective landing areas.

"Such pinpointing permits (1) precise control of distances in the Surveyor panoramas, with consequent accurate estimates of the dimensions of craters, rocks, etc.; (2) extrapolation of small-scale surface structure to other areas of similar appearance in Orbiter photographs; and (3) an independent check on landed-position coordinates determined from tracking data.

"The chosen landing point for Surveyor III was 43° W, 3° S, situated in a telescopically smooth mare area southeast of the crater Lansberg. This area had been photographed by Orbiter III shortly beforehand. An inspection of the Surveyor panoramas showed that the spacecraft had come to rest on the inner eastern slope of a crater, the angle of tilt being about 15°. From the disposition of several small craters and rocks [marked A, B, and C above], it was possible to draw a very rough plan of the neighborhood. Comparison of this plan with the Orbiter III photographs eventually allowed the pinpointing of the Surveyor III location.

"The photograph opposite is an enlargement of an Orbiter III high-resolution frame of the Surveyor III landing-point neighborhood [with craters A and C, and rocks B, marked]. The small white triangle represents the area enclosed by the spacecraft's foot-pads, and is correctly oriented. The rim of the 200-meter-diameter 'soft' crater defines Surveyor's horizon."

Surveyor V
Lands and Slides

Surveyor V, first of its spacecraft family to obtain information about the chemical nature of the Moon's surface, landed in Mare Tranquillitatis on September 11, 1967. The 50-picture mosaic of its landing site [above] is described in this way by R. J. DANKANYIN,

Manager, Surveyor Scientific Payload Systems, Hughes Aircraft Company:

"The spacecraft is on a 20° slope in a crater about 11 meters long, 9 meters wide, and 1.5 meters deep. The top edge of the crater is clearly discernible

about one-third of the way down from the top of the picture. The camera is actually about one-half meter above the edge of the crater, so the picture appears as though one were standing in the crater, looking toward and over the far side. Most of the small craters nearby are a few meters wide. The lunar horizon is about a kilometer away."

The spacecraft slid about 3 feet in landing, and the photograph its camera took of the slide mark [above] yielded useful information. FRANK B. SPER-LING, of Jet Propulsion Laboratory, said of it:

"The outstanding feature in this picture is the surface trench formed by the footpad during the landing process. The footpad first impacted just to the left of the rock, or clump, shown near the right margin. It penetrated the surface by approximately 4 inches while sliding 3 to 4 inches downhill. The spacecraft then rebounded to a height of 4 to 5 inches, still moving farther downslope. The bright area to the right of the leg marks the footpad's reimpact, with little penetration because of the greatly reduced impact energy. From there it slid to its final position.

"Lunar material was thrown out during this process, and some was deposited on top of the footpad. Also, some material crumbled from the walls of the first footpad imprint, obscuring its flattened bottom.

"Scientifically this picture is significant because it facilitates estimates of the frictional forces that acted between the footpad and the lunar surface."

SURVEYOR V

One high-quality Surveyor V photo [below], as JAMES D. CLOUD, Assistant Surveyor Program Manager, Hughes Aircraft Co., pointed out, shows how "the camera was used for visual engineering evaluation of the spacecraft's condition as well as for scientific purposes. Visible [upper left] is part of the small crater, 8 inches in diameter, 1 inch deep, formed when the vernier engines were static fired. Also visible are the large, spherical helium tank [top], the smaller nitrogen tank [with vertical black band], and the alpha-scattering electronic compartment [far left]."

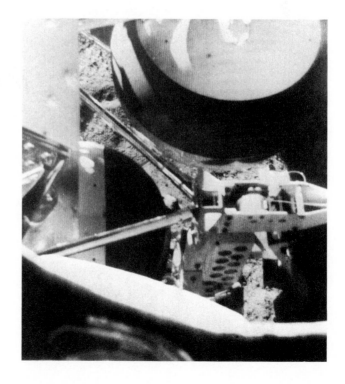

Investigating What the Moon Is Made of

"Surveyor V carried an instrument to determine the principal chemical elements of the lunar-surface material," explained ANTHONY TURKEVICH, Enrico Fermi Institute and Chemistry Department, University of Chicago. "After landing, upon command from Earth, the instrument was lowered by a nylon cord to the surface of the Moon [left, above]. The results of the measurements were telemetered back to Earth.

"Both photographs above were taken by Surveyor V's television camera. The right-hand one shows the chemical analyzer slightly displaced by the restarting of the spacecraft's vernier engines, 2 days after landing. The displacement enabled it to analyze a second sample of lunar-surface material.

"The instrument under discussion is a cubical box, approximately 6 inches on a side, sitting on a 12-inch white plate. The box has an opening in the bottom through which alpha particles from the radioactive element curium, provided by the Argonne National Laboratory of the Atomic Energy Commission, are directed toward the lunar surface underneath. A

few of these alpha particles are scattered back; some of them produce protons by nuclear reactions in the lunar material. From the number and energy of the scattered alpha particles and protons, the chemical composition of the lunar surface can be deduced.

"In this way," Turkevich continued, "it was determined that the most abundant elements at the mare site where Surveyor V landed were oxygen, silicon, and aluminum, in decreasing order. This is the order of abundance of the chemical elements on the crust of the Earth. The relative amounts of the elements resemble those in terrestrial basalts. This similarity suggests that the lunar material on the surface of the mare has been subjected to a geochemical process of differentiation similar to that undergone by the material of the Earth's crust."

Surveyor V was carrying a magnet on footpad 2 [top left, opposite page] when it slid downslope upon landing. JANE NEGUS DE WYS, of Jet Propulsion Laboratory, commented on the picture as follows:

"The permanent magnet [500-gauss strength] is at left in the photo, the unmagnetized control bar at right. A relatively small amount of iron material is seen adhering to the magnetic pole faces.

"Laboratory studies of impact in powders of different rock types and impact in powdered basalt with additions of powdered iron in varying amounts were used to interpret the pictures of the magnet that landed on the Moon. The closest resemblance was to the finely powdered basalt with no addition of powdered iron. This was in agreement with the alpha-scattering data. Since the magnet and alpha-scattering data indicate basalt, the lunar surface material in this area is of volcanic origin.

"The two general theories of lunar-crater formation are (1) meteoritic bombardment and (2) volcanic or gaseous emission.

"If all lunar craters and pulverization of the surface material were due to meteoritic impact," Mrs. de Wys concluded, "some addition of magnetic iron to the lunar-surface material would be expected. The magnet on the Moon does not indicate such an addition. Therefore, the extent of meteoritic bombardment and its role in lunar-surface morphology should perhaps be reexamined in the light of this evidence."

The dark semicircle in the lunar surface, located in the center of the photograph above, is a small, man-made crater, caused by a brief pulse of exhaust from one vernier engine. It was the result of a test that Surveyor performed in behalf of the manned lunar-landing program. JAMES H. TURNOCK, Apollo Program, NASA, wrote:

"This successful experiment was performed at the request of the Apollo Program to determine by extrapolation what the effects on the lunar surface will be from the exhaust of the lunar module's descent engine. We wondered whether large clouds of dust would be raised, which would impair the astronauts' visibility. This information could not be obtained during a Surveyor landing, since the engines are cut off about 14 feet above the surface.

"Therefore, on September 13, 1967, about 53 hours after Surveyor V's landing, its engines were ignited and allowed to burn for 0.6 second. The resulting crater formed beneath engine 3, shown here, is about 8 inches in diameter and a little over an inch deep. That is about what we expected, and indicates that the engine-exhaust effects on the lunar surface should be no problem for Apollo. This, therefore, represents another important data point that the Surveyor Program provided for Apollo."

Surveyor VI Checks Out

Another Landing Site for Apollo

Surveyor VI landed on the flat surface of Sinus Medii, at 1.40° W, 0.49° N, on November 10, 1967, at 01:01:05 GMT. It came to rest near a mare ridge, visible in the lunar vista above.

In the course of its mission, this spacecraft took more than 30 000 photographs. It also took them from two slightly differing perspectives. At one point, its vernier engines were restarted and it was made to hop 8 feet. It then took fresh measurements and pictures from its new location (opposite page).

Though Surveyor VI and its three successful predecessors landed at lunar locations that were thousands of kilometers apart, the sites chosen for them to investigate proved to be strikingly alike, and promising for manned landings.

"The mosaic at the top of this page is the first view of a lunar wrinkle ridge from the surface of the Moon," declared JOHN B. ADAMS, of Jet Propulsion Laboratory. "Wrinkle ridges are low, sinuous features on the maria that have been observed for many years through telescopes, and, more recently, in Lunar Orbiter photographs.

"In this view from Surveyor VI, in Sinus Medii, the ridge extends northeast along the horizon. Each frame in the mosaic covers a field of view of 6°. Notable features are the gentle slopes along the ridge flank facing the camera and the abundance of rock blocks on the ridge relative to the mare material near the spacecraft [foreground]. Blocks up to 1 meter in diameter are prominent around the two subdued craters on the ridge [right of center].

"From this photo and others," said Adams, "it is thought that craters that were formed on the ridge excavated more solid rock than did similar-sized craters in the flat mare. This suggests a thinner mantle of particulate rock on the ridge. The origin of the ridge is still a subject of debate, although wrinkle ridges in general are thought to have formed by deformation of the mare material.

"The terrain photographed by Surveyor VI in Sinus Medii does not appear to be excessively hazardous for landing spacecraft. Although the region is highly cratered, most slopes are gentle, and there are few rocks—compared, for example, to the rock-strewn terrain in Oceanus Procellarum seen by Surveyor I. Sinus Medii is one of several prime sites for future manned landings."

After having been commanded to hop from its original landing position, Surveyor VI took additional pictures, of which the scene above is one.

"This view of the lunar surface to the southeast of Surveyor VI was taken shortly before sunset," explained JACK N. LINDSLEY, of the Jet Propulsion Laboratory. "Long shadows cast by the late-afternoon Sun accentuate the irregularities of the Moon's surface in the Sinus Medii area. Although there are many craters and depressions, which appear as dark areas because they are filled with shadows, due to the low Sun angle, there are very few rocks of significant size. The shadow that appears on the bottom edge of the picture about 2 inches inward from the lower-left corner is caused by the rim of a depression left by one of the spacecraft's feet when it first landed.

"The nearest area shown in the picture is about 12 feet from the camera, and the horizon is about one-half mile distant. A large crater is located near the horizon, to the right of the picture's center.

"The picture is a mosaic made from 20 individual wide-angle pictures taken by Surveyor VI's television camera, and covers 120° of the panorama. These 20 pictures are part of a series of 120 pictures taken to provide complete, 360° coverage of all the viewable lunar surface and of the spacecraft structure as well. During the sequence of picture taking, pictures were televised from the Moon to Earth at the rate of one about every 5 seconds. Each picture required approximately five commands to be radioed from Earth to the spacecraft in order to position the camera and snap its shutter."

Lessons in the
Lunar Soil

"The photograph on the opposite page was taken by Surveyor VI on November 17, 1967, after the first liftoff-and-translation maneuver on the Moon," explained Benjamin Milwitzky, Surveyor Program Manager, NASA. [That is true as well of the right-hand panel of the illustration at the bottom of this page.] "In this historic operation, the spacecraft was moved to a new location, approximately 8 feet from its original landing point.

"The picture shows the effects of the vernier-rocket-engine blast on the double imprint previously made in the lunar surface by one of the spacecraft's crushable blocks during the initial touchdown. Material has been excavated from the imprints and deposited over a broad area on the lunar surface in a thin, dark coating forming a sheet-and-ray pattern. As has been observed in all Surveyor missions, the underlying lunar material is appreciably darker than the surface material. The latter appears to be a very thin layer whose higher reflectivity may result from the effects of solar and cosmic radiation.

"From this and other Surveyor VI photos, it appears that rocket exhaust gases can strongly erode and excavate an area that has been previously disturbed, but have relatively little effect on the undisturbed lunar surface. These observations further reinforce the concept that the lunar surface is composed of very fine particles lightly held together by vacuum cohesion resulting from long exposure to the near-perfect void of outer space."

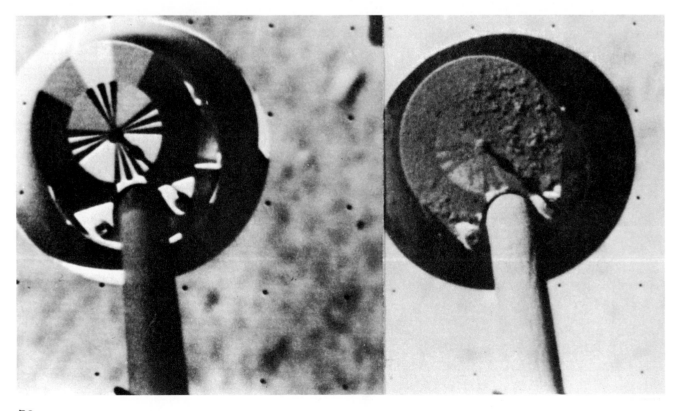

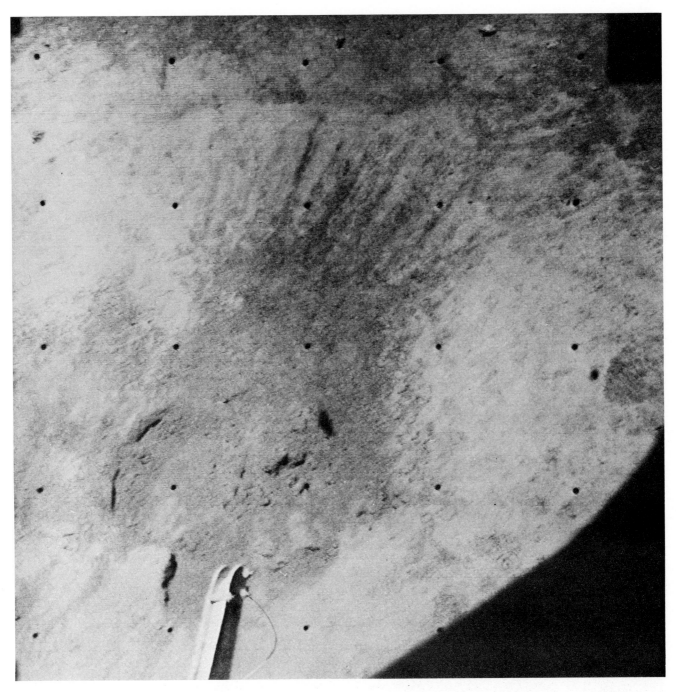

The pictures at left were taken before and after the vernier-engine firing. "Prior to the hop, the photometric chart on the end of omniantenna B was clean," wrote Raoul Choate, Lunar and Planetary Sciences, Jet Propulsion Laboratory. "After the hop, the chart was covered with a coating of lunar soil up to 0.9 millimeter thick. Because the rest of the spacecraft remained relatively clean after the firing, it is surmised that the chart was hit by an individual soil clump. The clump was probably 2 to 3 inches in diameter.

"This photograph, better than any other Surveyor picture, illustrates the ability of lunar soil to adhere to vertical and near-vertical faces."

Surveyor VII Examines Tycho's Highlands

For scientists seeking insight into the Moon's structure and history, Surveyor VII obtained the panoramic picture on the next page in the area that is shown below it in an Orbiter V photograph.

"The Surveyor Program," ROBERT C. SEAMANS, JR., then NASA's Deputy Administrator, wrote at its conclusion, "was initiated in 1960 before a national commitment was made to land astronauts on the Moon and return them safely. Hence the objective was initially to land spacecraft on the Moon in order to obtain maximum scientific information after touchdown. The Surveyor operation differed from Ranger, which obtained data prior to impact, and Orbiter, which photographed the lunar surface while orbiting the Moon. The Surveyor program objectives gradually shifted emphasis as the Apollo manned lunar-landing program became more firm. The designers of the Apollo lunar lander had to know the bearing strength of the Moon's surface in order to establish requirements for impact velocity and footpad area. Many other questions arose that could only be answered by landing Surveyor on possible Apollo landing sites. At the time these missions were planned, it was anticipated that the entire seven Surveyor spacecraft would be needed to investigate several Apollo sites. Actually four successful landings of Surveyor were achieved with the first six spacecraft and all Apollo requirements were then satisfied. It should be recognized that an enormous amount of valuable scientific information was obtained from each of these four missions, but the landing locations were restricted to the nearly equatorial sites achievable with Apollo. The success of these missions permitted the seventh spacecraft to be sent to a location where the maximum additional scientific data could be obtained. The highlands about 18 miles north of the large crater Tycho were selected. Many thousands of useful photographs of the Moon were obtained by Surveyor VII television cameras from this site. A selection of these pictures was used to form the panoramic view of the lunar terrain northeast of the spacecraft (on the top of the facing page).

"In this illustration the distance from Surveyor to the horizon varies from hill to hill, but the horizon in the center of the picture is about 8 miles away. Near the horizon a series of ridges and hills can be seen somewhat similar in form to many rounded hills found in parts of the eastern United States. Just below the horizon are small ravines and gullies probably formed during the flow of debris at the time or shortly after it was deposited. In the foreground is a rocky crater which lies about 18 feet from Surveyor's camera. The crater, about 5 feet in diameter, is filled with rocks. The rock on the near rim of the crater is 2 feet across and casts a 4-foot-long shadow to its left. The photo interpretation was made by Dr. Eugene Shoemaker of the U.S. Geological Survey, who was the principal investigator for the Surveyor television experiment.

"The Surveyor VII mission also permitted other than photographic data to be obtained. For example, as in previous missions, an alpha-backscatter experimental package developed by Dr. Anthony Turkevich of the University of Chicago was lowered to the lunar surface. This experiment provided data on the chemical composition of the lunar surface."

The Orbiter V high-resolution photo (below on the facing page) shows where Surveyor VII landed.

"This region was chosen," explained ABE SILVERSTEIN, Director, Lewis Research Center, NASA, "because it is believed to be covered by material ejected from beneath the lunar highlands when Tycho was formed. Because of the roughness of the ejecta blanket, Surveyor VII's target area was a circle only 12.4 miles in diameter, one-third the normal size, in order to provide a reasonable probability for a survivable landing. Both the launch vehicle and the spacecraft performed so precisely that, after a single midcourse maneuver, Surveyor VII landed less than 1½ miles from the center of the small target circle.

"The landing point on the photo was determined by Dr. Ewen Whitaker of the University of Arizona by correlating more than 25 lunar-surface features visible in both Orbiter V and Surveyor VII pictures.

"This is another good illustration of the manner in which the Surveyor and Orbiter spacecraft were used in combination for effective lunar exploration."

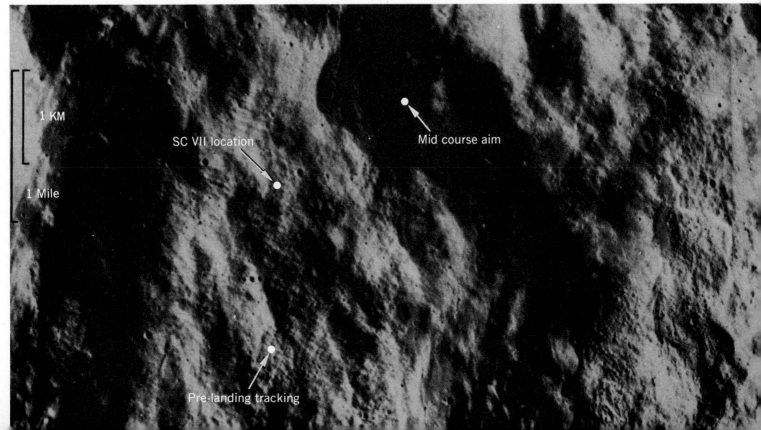

1 KM

1 Mile

Mid course aim

SC VII location

Pre-landing tracking

Spotting Two
Laser Beams
From Earth

Surveyor VII, launched on January 7, 1968, at 06:30 GMT, successfully landed on the Moon on the Tycho ejecta blanket on January 10, 1968, at 01:05 GMT. The coordinates of the landing site were 40.89° S latitude and 11.44° W longitude. From that position, it transmitted 21 274 television pictures during the first lunar-day operations. Its camera system also spotted laser beams aimed toward it from two observatories on Earth.

The Surveyor photograph of the partially sunlit Earth [top left], is described by C. O. Alley, of the University of Maryland: "Surveyor VII's TV camera detected as starlike images two narrow laser beams sent to the Moon from, respectively, the Kitt Peak National Observatory, near Tucson, Ariz., and the Table Mountain Observatory, near Los Angeles.

"The blue-green argon-ion laser beams seen within the white circle on the photograph each contained only about 1 watt of power," Alley explained, "but appeared somewhat brighter than the brightest star, Sirius.

"This engineering test of the aiming of the beams, a few miles wide at the Surveyor site, was conceived and coordinated by me and my fellow-professor D. G. Currie, of the University of Maryland's Department of Physics and Astronomy, to gain experience for the Apollo laser-ranging retroreflector experiment, for which I am principal investigator."

"The photograph at bottom left shows the detail of the center part of a rocky crater at the Surveyor VII landing site," reported R. H. Steinbacher, of Jet Propulsion Laboratory. "The narrow-angle picture of rocks 18 to 20 feet from the spacecraft has a field of view of about 3 feet diagonally across the picture. Some of the rocks are as much as 1 foot in size. The blocky, angular appearance of the rocks is characteristic of what is believed to be a relatively new crater. Close examination of these rocks and of the many others at this highland landing site was a main scientific objective of this mission."

SURVEYOR VII

The attention-commanding photograph at the right was among those that Surveyor VII took of lunar rocks of special geological interest.

"This remarkably detailed, narrow-angle photograph of a cracked lunar rock gave us our first glimpse of fissure cracks in the blocky material of the Moon's surface," declared GLENN A. REIFF, Manager, Mariner IV and V Programs, NASA. "The cracks in view here are about one-fourth inch wide, and were doubtless caused in place by expansion and contraction in the very severe temperature extremes of the airless Moon.

"These rocks, approximately 6 inches in size, are dense and fine-grained. Differences in light reflectivity on their surfaces suggest that the rocks contain at least two types of minerals. That curious image in the lower left-hand corner is a reflection from part of the spacecraft structure.

"I'm looking forward to the day," Reiff continued, "when we can obtain photographs of such exceptional detail as this from the surface of Mars or one of the other planets. However, we can't anticipate obtaining Martian surface detail comparable to that of Surveyor VII's pictures of the Moon until such time as spacecraft actually land on the surface of the Red Planet.

"What we can look forward to, though, is that a number of the principles pioneered by Surveyor, such as guidance and landing techniques, can be used in future Mariners for exploring Mars."

"The narrow-angle picture at bottom right, taken by Surveyor VII's television camera, shows the rolling lunar terrain northeast of the landing site," wrote F. N. SCHMIDT, of Bellcomm, Inc. "Surveyor VII landed about 18 miles north of the large crater Tycho, the most prominent of all lunar craters at full Moon. The undulating surface is part of a series of ridges and hills that are a characteristic feature on the flanks of most large lunar crater rims. They were probably formed by the flow of debris ejected by the meteor impact that excavated the crater."

SURVEYOR VII

Recording Details
of Lunar Highlands

Surveyor VII

Repairs Itself . . .

"The alpha-scattering instrument's sensor head is shown on the left-hand side of this Surveyor VII picture [top left]," explained C. E. CHANDLER, Surveyor Project engineer, Jet Propulsion Laboratory. "Because of a problem in the mechanical deployment mechanism, the sensor head did not drop to the surface on command but remained, as shown, in its background-count position. The scoop of the surface sampler, shown in the middle of the photograph, was then manipulated by command in an attempt to nudge the head and free the deployment mechanism. This initial effort did not succeed, but later the head was successfully deployed by a similar attempt, and the alpha-scattering instrument obtained lunar-composition data from a total of three sites close to the spacecraft." The instrument, in fact, spent a total of 43 hours 5 minutes in surface analysis; 12 hours 7 minutes in background count.

One of the several objectives of the Surveyor VII mission was to manipulate the lunar material with the surface sampler in view of the television camera. Six bearing-strength tests were conducted by pressing the bucket of the surface sampler against the lunar surface while recording drive-motor currents. Also, the sampler dug several trenches, one of which (bottom left) was approximately 16 inches long and a little more than 6 inches deep.

In commenting on that photograph, RONALD F. SCOTT, of the California Institute of Technology, wrote: "This picture shows the second trench dug by Surveyor VII's surface sampler. The scoop of the sampler, with its serial number, can be seen at the bottom of the picture, after completing a pass through the trench. The cohesive nature of the lunar soil is clearly indicated by the very smooth vertical wall left by the sampler scoop.

"The sampling device was designed to perform tests such as this on the lunar surface to determine the mechanical properties of the surface material," Scott continued. "However, Floyd Roberson, my JPL colleague, and I, who operated the sampler, also used it on Surveyor VII to move the sensor head of the alpha-scattering experiment down to the lunar surface. Later, we used it to move the head [see facing page] to different positions on the surface."

Still another use was found for the versatile sampler. When temperatures in the sensor head rose too high, the sampler scoop was used like a parasol, shading the instrument from the Sun.

SURVEYOR VII

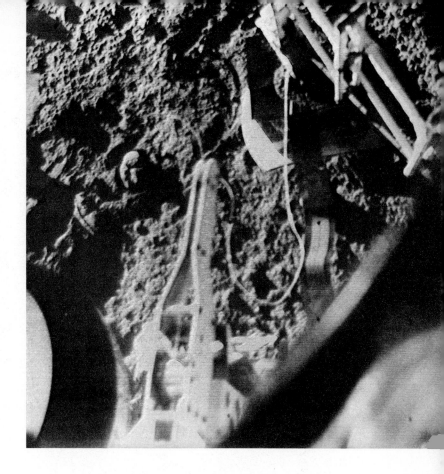

"The 3-inch rock at left center in the picture at top right was the second of three samples analyzed by means of the alpha-scattering experiment on Surveyor VII," said ERNEST J. FRANZGROTE, of Jet Propulsion Laboratory. "The picture was taken as the surface-sampler mechanism [upper right] was moving the alpha-scattering instrument to its third location, a plowed-up area at the right-hand edge of the picture. The circular impression on the lunar surface surrounding the rock shows the resting place of the bottom of the instrument during the analysis."

Surveyor data have shown the lunar maria to be remarkably uniform in their physical and chemical properties. Chemical analysis in the lunar highlands, by the crater Tycho, differed in that the concentration of iron and neighboring heavy elements was only about half that found in the maria. This deficit may explain the higher albedo (greater brightness) of the highlands. It also suggests that the highland material may be of somewhat lower density.

"This picture at bottom right, taken by Surveyor VII's 600-line TV camera, consists of two views of the lunar surface," explained D. H. LE CROISSETTE, Manager, Surveyor Instrument Development, Jet Propulsion Laboratory. "The central part of the picture shows the surface sampler poised above the lunar surface, as viewed through an auxiliary mirror placed on the mast of the spacecraft. The mirror was 3½ by 9½ inches in size, and was positioned so that a stereo view of a portion of the surface could be obtained where the surface sampler could reach.

"In this picture, the sampler had already conducted a bearing test in the soil and had started to dig a trench. The rest of the photograph is a direct view of another part of the lunar surface, looking between the mirror and some structural members of the spacecraft. The rocks shown in this view are about 1 foot across."

The hard work of Surveyor VII's long lunar day, on which the Sun set on January 22, 1968, also included collecting data on touchdown dynamics, temperatures, and radar reflectivity. The spacecraft obtained data on touchdown dynamics by means of telemetry from strain gages mounted on its landing legs; radar-reflectivity data in the course of its terminal descent; and thermal data from onboard sensors.

SURVEYOR VII

and Photographs
a Hard Day's Work

Man's First Look at Earth From the Moon

Describing the spectacular, historic view above, FLOYD L. THOMPSON, then Director, Langley Research Center, wrote: "At 16:35 GMT on August 23, 1966, the versatile manmade Lunar Orbiter spacecraft responded to a series of commands sent to it from Earth, across a quarter-million miles of space, and made this over-the-shoulder view of its home planet from a vantage point 730 miles above the far side of the Moon.

"At that moment," Thompson continued, "the Sun was setting along an arc extending from England [on the right] to Antarctica [on the left]. Above that line, the world, with the east coast of the United States at the top, was still bathed in afternoon sunlight. Below, the major portion of the African Continent and the Indian Ocean were shrouded in the darkness of evening.

"By this reversal of viewpoint, we here on the

84

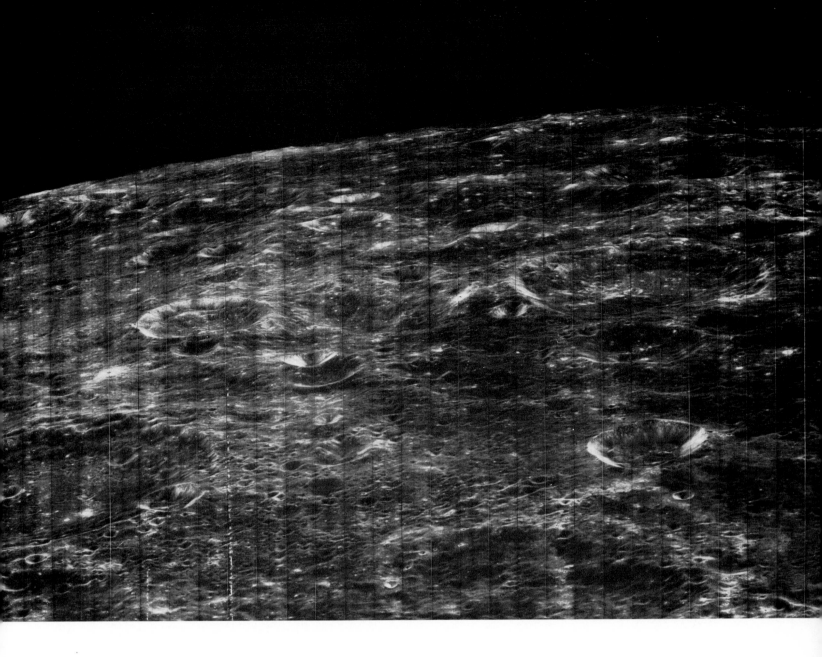

. . . and an Oblique View of the Moon Itself

Earth have been provided a sobering glimpse of the spectacle of our own planet as it will be seen by a few of our generation in their pursuit of the manned exploration of space. We have achieved the ability to contemplate ourselves from afar and thus, in a measure, accomplish the wish expressed by Robert Burns: 'To see oursels as ithers see us!' "

Also visible in dramatic new perspective in this photograph is the singularly bleak lunar landscape, its tortured features evidently hammered out by a cosmic bombardment that may have extended over billions of years.

Because the airless, weatherless Moon appears to preserve its surface materials so well, it may serve science as an illuminating record of past events in the solar system. ROBERT JASTROW, Director, Goddard Institute for Space Studies, has called the Moon "the Rosetta Stone of the planets."

Orbital Tours
Bring Surprises

Orbiter missions have been critically important in selecting lunar sites most suitable for astronaut landings. They have also provided what some scientists have called a "truly staggering" amount of new information about the Moon's surface.

JOHN F. McCAULEY, U.S. Geological Survey, wrote of the photo at the right: "This fascinating crater on the far side of the Moon has a gently domed floor, crisscrossed with linear depressions [about 1 kilometer wide] that give it the appearance of a turtle's back. Similar, crudely polygonal fracture patterns occur on Earth in domes formed by upward movement of plastic or liquid material.

"The doming of the floor of this crater could be the product of local magmatic activity or of slow isostatic rebound. Both alternatives point to probable activity of internal origin in the lunar crust."

Regarding the photo below, DONALD E. GAULT, Ames Research Center, declared: "Here is a graphic example on a colossal scale of one of the basic processes active on the lunar surface. Of these two unnamed craters on the far side of the Moon, the smaller, 45 kilometers in diameter, is the older,

since its northwest rim has collapsed and been displaced violently inward by the formation of the larger crater [55 kilometers in diameter]. This inward avalanche of rocky debris from the collapsed rim, together with ejecta spewed out along ballistic trajectories from the larger crater, has partially filled and obliterated the smaller structure. Thus, the mechanism by which impact craters are produced is an equally efficient agent for destroying craters, and impact scars do not necessarily provide a complete record of cratering events."

"From the best Earth-based telescopic photographs of the Moon, we selected eight areas across the Equator that appeared to be sufficiently free of craters to permit the astronauts to land safely," explained URNER LIDDEL, of NASA. "When Lunar Orbiter I began photographing these areas, with between 100 and 1000 times the clarity of telescopic pictures from Earth, its high-resolution camera did not work dependably. Most of its pictures, therefore, were of medium resolution, like the one on the opposite page. Even these, however, revealed that lunar-mare areas that looked smooth from Earth were actually covered with small craters."

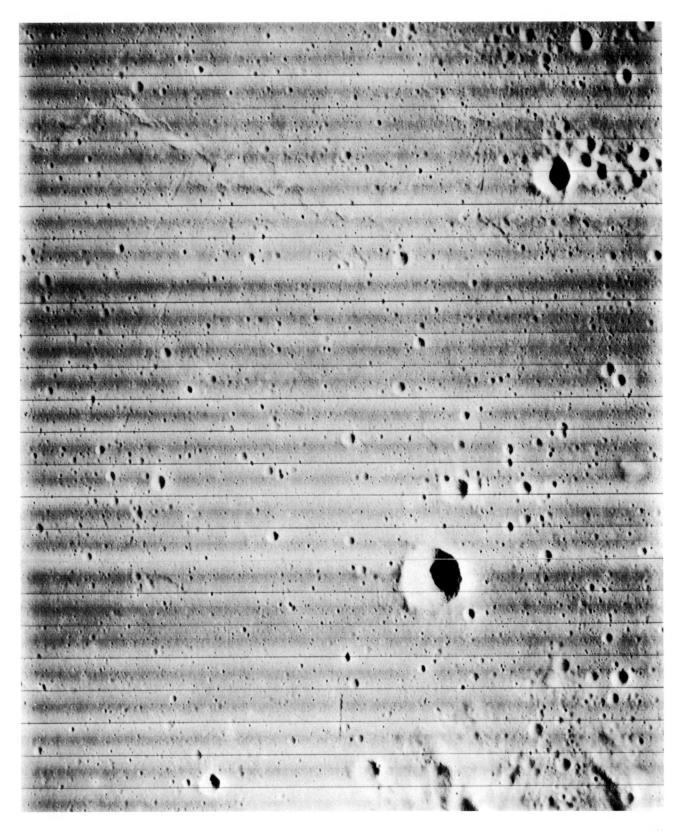

ORBITER I

Orbiter II Takes the "Picture of the Year"

The breathtaking lunar landscape displayed on the opposite page was one of Orbiter II's grandest photographic achievements. [Compare it with one of the best pictures of the same scene that Earth can offer (right), made with the 61-inch telescope of the Lunar and Planetary Laboratory, University of Arizona.]

"On first seeing this oblique view of the crater Copernicus," declared ORAN W. NICKS, Deputy Associate Administrator, Office of Space Science and Applications, NASA, "I was awed by the sudden realization that this prominent lunar feature I have often viewed by telescope is a landscape of real mountains and valleys, obviously fashioned by tremendous forces of nature. It is no wonder that some writers immediately classified it as the 'Picture of the Year'! [Some, with understandable enthusiasm, even hailed it as the 'Picture of the Century.']

"Lunar Orbiter II recorded this image at 7:05 p.m. e.s.t. on November 28, 1966, from 28.4 miles above the Moon's surface, and about 150 miles due south of Copernicus. The clarity of the view is attributable to the absence of atmosphere. A photograph from similar altitudes of distant features on Earth would never be as sharp, because of haze.

"Copernicus is about 60 miles across and 2 miles deep; 3000-foot cliffs, apparently landslide scarps, can be seen. Peaks near the center of the crater form a small mountain range, about 1500–2000 feet high and 10 miles long.

"The Lunar Orbiter photography was accomplished with two cameras," Nicks explained, "one having a 3-inch focal length and the other a 24-inch focal length. These cameras were boresighted, so that each high-resolution photo was always contained in a moderate-resolution frame. Lunar Orbiter cameras were relatively conventional film cameras that combined a Bimat chemical development process with an electronic scanning readout for transmission by radio to Earth. The film images provided a very effective method of storing information for transmission bit by bit, at a modest rate."

"The telescopic view of Copernicus shown above is one of the finest photographs ever taken of this region from the Earth, and shows features as small as 2500 feet across," said WILLIAM E. BRUNK, Planetary Astronomy Chief, Lunar and Planetary Programs Directorate, NASA. "It is not possible to photograph smaller features because of the turbulence in the Earth's atmosphere.

"The crater Copernicus, a prominent feature on the lunar landscape, is believed to have resulted from an impact of a second body with the Moon," Brunk continued. "The 'keyhole'-shaped crater,

Fauth, is seen at the bottom of the photograph; the Carpathian Mountains at the top. Characteristics of the landscape are clearly shown by the shadows produced by the rising Sun, whose elevation was approximately 10 degrees above the horizon. Numer-ous mounds are visible on the floor of Copernicus, in addition to the central peaks."

The reader will find it interesting at this point to compare the pictures shown here with Orbiter V's vertical views of Copernicus on pages 116–117.

Ancient Lava Flows
and a Rock Field

Evidence of apparent volcanic activity in the Moon's past is seen in the top photo here. CLIFFORD H. NELSON, Lunar Orbiter Project Manager, Langley Research Center, NASA, described this view:

"This magnificent northerly oblique photo from Lunar Orbiter II brings to view the Marius Hills region shortly after lunar sunrise. The picture was taken on the last day of Lunar Orbiter II photography, November 25, 1966, with image-motion compensation turned off.

"The slow curvature of the horizon is 250 miles from Orbiter's wide-angle (40°) camera. The crater Marius, at 50.7° W, 12° N, is about 25 miles in diameter and 1 mile deep. Its floor is surprisingly level.

"The central portion of this scene is dominated by a spectacular array of dome structures. These domes are up to 10 miles in diameter and as much as 1500 feet high. Many features are similar in appearance to volcanic domes in northern California and Oregon. They are interpreted to be the result of upward movement of magma that has warped the overlying rock and in some cases spilled out on the surface as lava.

"The irregular lines of hills that cross this lunar scene diagonally appear as wavelets washing across shallow beach areas on Earth. These hills have apparently been formed from lava flowing out through cracks in the Moon's crust."

Some of the largest rocks seen in early lunar missions are visible in the Orbiter II photograph on the opposite page.

"This rock-strewn area," explained FRED A. ZIHL-MAN, Lunar and Planetary Programs, NASA, "is situated in the southeastern part of Mare Tranquillitatis. Some of the larger rocks are about 30 feet across. This photograph, covering an area 1200 by 1500 feet, is enlarged five times from the original filmstrips.

"The photograph is interesting because of the relatively large size of the rocks shown and because of the circular distribution of rocks in the center of the picture. Some lunar scientists suggest that the rocks were broken and excavated from subsurface layers by meteoric impacts, and that in this area, which is part of a ridge, finer ejecta material has been gradually removed by some 'mass-wasting' process that uncovered the rocks in the process of filling the impact craters.

"Distributions of rocks were also photographed by Surveyors I and III in the region of Oceanus Procellarum. Surveyor I pictures show a fairly large number of blocks on the horizon, the coarser ones about 3 feet across. At the Surveyor III landing site, the coarsest blocks, scattered mainly in two distinct strewn fields, range in size from a few inches to approximately 6 feet across. The rocks shown here, considerably larger than those photographed by Surveyors I and III, may have been raised by larger meteoric impacts and from greater depths."

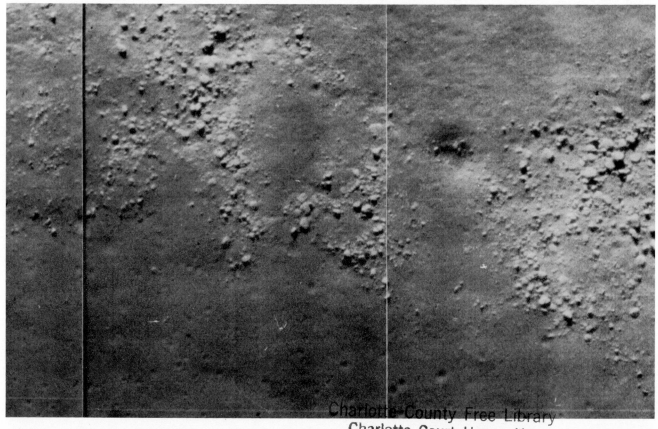

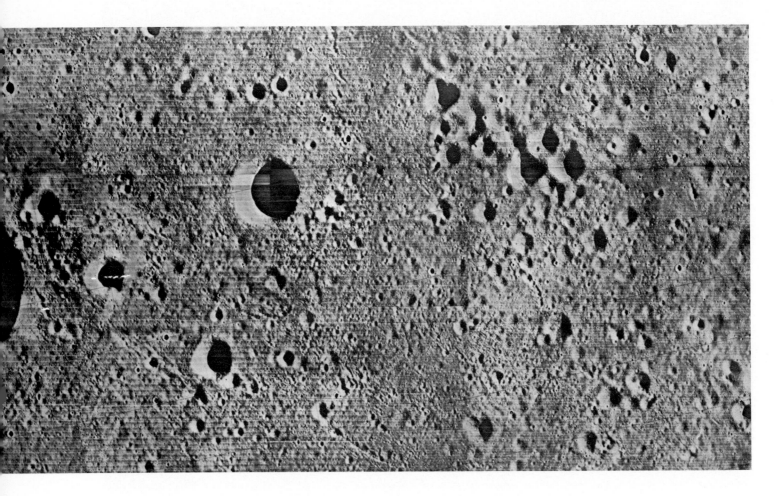

Orbiters Help Find a Place for Men To Land

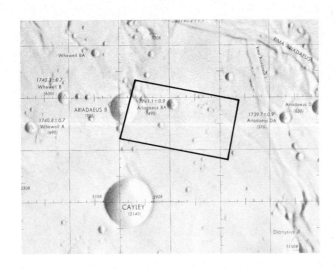

"We naturally began our selection of potential Apollo landing sites by examining Earth-based data," said ARTHUR T. STRICKLAND, Cartography Chief, Lunar and Planetary Programs, NASA, "and choosing generally smooth-looking lunar areas, ostensibly devoid of craters, as the most promising. The smooth area outlined by a black rectangle on a small portion of the best available topographic chart of the Moon [at left] is one area we originally picked. This site, approximately 20 miles long and 11 miles wide, is in the northeastern section of the central highlands, slightly south of the Rima Ariadaeus and west of Mare Tranquillitatis.

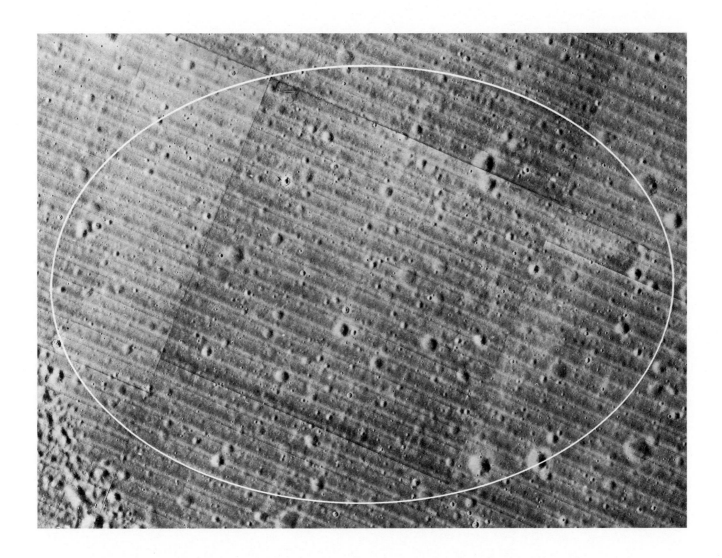

"Detailed photographic examination of this area by Lunar Orbiter II's camera, as shown at the top of the preceding page, demonstrated to us how inadequate Earth-based instruments and methods are to define the lunar terrain for a Lunar Module's landing."

"When Lunar Orbiter II took that high-resolution photograph of the site," commented SAMUEL C. PHILLIPS, Apollo Program Director, "the terrain was revealed to be extremely rough and obviously unsuitable for an early manned landing attempt.

"We were delighted, however, to find many areas similar to the one marked with a white ellipse above, which are smoother and will no doubt yield acceptable sites for manned landings. The area above, shown in high-resolution photography obtained from Lunar Orbiter III, is one of the likeliest candidates from which the initial Apollo landing site will be selected. The ellipse [a shape dictated by the nature of the Lunar Module's landing approach] covers approximately 5 miles by 3¼ miles. It is located in southeast Oceanus Procellarum, at 3°30′ S and 36°25′ W. Its selection was based on considerations of topography, surface bearing strength, and flight operations."

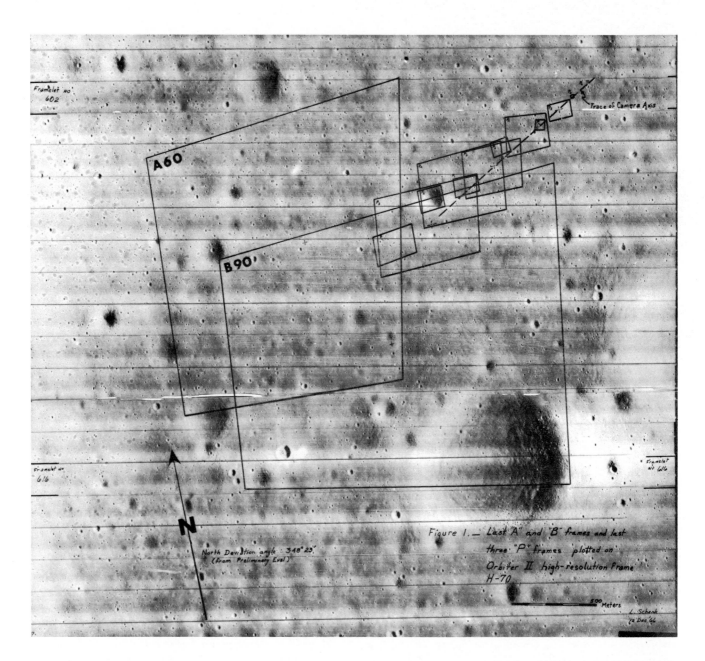

Framelet No 602

A60

B90

Trace of Camera Axis

Framelet No 616

Framelet No 616

N

North Deviation angle : 348° 25'
(from Preliminary Eval)

Figure 1. — Last "A" and "B" frames and last three "P" frames plotted on Orbiter II high-resolution frame H-70

500 Meters

L. Schenk
12 Dec 66

The Spot Where Ranger VIII Crashed

"In November 1966, Lunar Orbiter II took this high-resolution photograph of the area on the western edge of Mare Tranquillitatis where Ranger VIII had impacted on February 20, 1965," explained NEWTON W. CUNNINGHAM, former Ranger Program Manager, NASA. Ranger VIII had photographed parts of this area, and in this picture the mark it left on the Moon was found.

"Ranger VIII had not been able to photograph its impact point because the optical axis of its cameras was not alined with the trajectory of the incoming spacecraft. However, framed regions representing the last areas photographed by Ranger VIII's 'A,' 'B,' and 'P' cameras were superimposed on the

Lunar Orbiter II picture, as has been done here. By plotting the resultant camera trace, it was then possible to determine Ranger VIII's trajectory trace on the surface of the Moon.

"A 10–× enlargement [shown above] was next made of two framelets of Lunar Orbiter II's photo [each covering 200 meters of surface].

"On this enlargement," Mr. Cunningham continued, "the extended plot of Ranger VIII's trajectory crossed from lower left to upper right, passing near the two craters C1 and C2. On the basis of Ranger VIII's mass of 800 pounds, its impact velocity of 2.653 km/sec, and the computed angle of approach from the local horizontal (41°), it was concluded that C1 could be the crater made by Ranger VIII's impact. Additional support for this conclusion comes from the bright halo of new material thrown up from the surface and the generally elliptical shape of C1, pointing along the direction of Ranger VIII's flight."

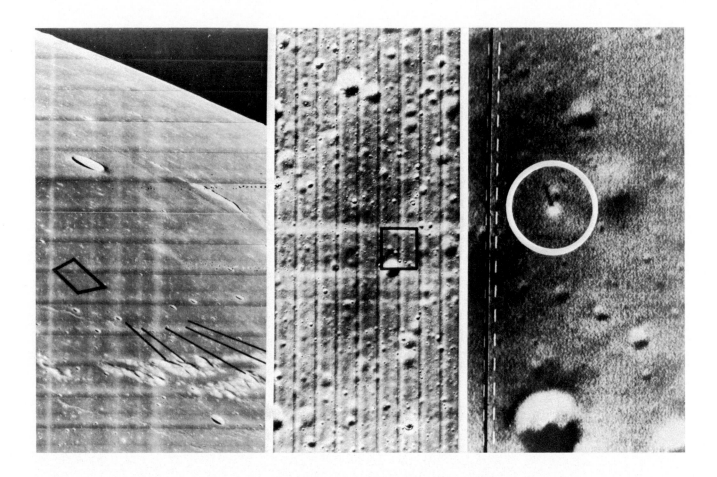

Orbiter III Locates Surveyor I

"These photographs of Surveyor I's landing area in Oceanus Procellarum were taken by Lunar Orbiter III on February 22, 1967, from an altitude of 30 miles," explained WALTER JAKOBOWSKI, Surveyor Program engineer, NASA. "The left-hand panel is an oblique, wide-angle view showing a partial ring structure of low mountains, with the crater Flamsteed at the southern tip of the ring. By triangulating from those mountains, which were visible in Surveyor I's photographs, the landed spacecraft's approximate location was determined. It was then possible to find Surveyor I in the center picture, which is Orbiter III's high-resolution photograph of the area. The right-hand panel is a portion of the center picture enlarged by a factor of 8. The space-

craft, with its three bright landing-gear surfaces, appears as a triangular white spot in the center of the circle. To the top and left of this spot is the 30-foot shadow cast by Surveyor's solar panel and antenna.

"The locating of Surveyor I in these pictures is significant in that Orbiter's vertical view permitted the scaling of distances and determination of the size of features seen in the Surveyor photographs, which otherwise would not have been possible. It also will permit extrapolation of Surveyor's detailed findings on local lunar-surface characteristics to other areas of the Moon that, according to Orbiter pictures, are similar in general characteristics to the area in which Surveyor I landed."

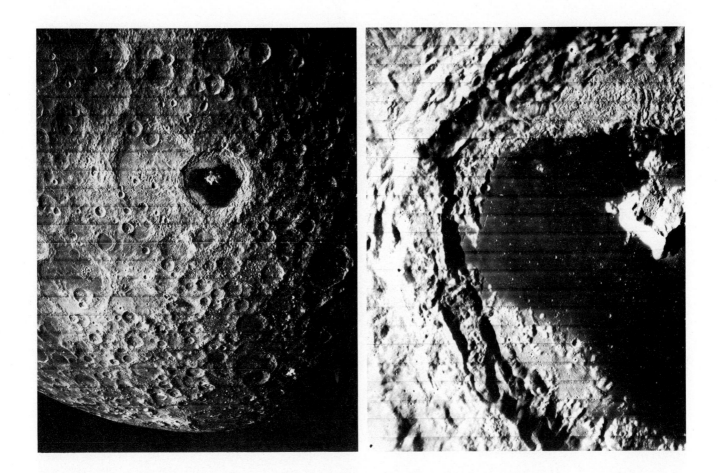

. . . and Peers Into Tsiolkovsky

"Here," explained ISRAEL TABACK, Lunar Orbiter Project Office, Langley Research Center, NASA, "are a wide-angle [above, left] and a telephoto photograph of the far side of the Moon, taken simultaneously by Lunar Orbiter III on February 19, 1967. The spacecraft was in a position near the Equator and about 900 miles above the surface. It was commanded to look obliquely south. The area of the far side included in the wide-angle photo is the eastern part of the Southern Hemisphere and spans from about 10° S, at the top, to 55° S, at the bottom. The width of the area covered near the Equator is approximately 750 miles.

"The telephoto photograph shows the central region of the area covered by the wide-angle photo-graph, at a much better resolution. Each strip [delineated by faint parallel lines] is about 4¾ miles wide.

"The area shown in the wide-angle frame is predominantly cratered, as is the entire far side. In this way it differs considerably from the front side, where numerous mare basins are present. There are some of these on the far side, and the crater with the very dark floor, seen near the center of the wide-angle photo, is a prime example [compare p. 45]. This crater, partially covered by the telephoto frame [right] at a significantly better resolution, is about 150 miles in diameter. It has been named Tsiol-kovsky. Its features seem indicative of the processes that occur during mare formation."

The Moon's Face Grows More Familiar

"The most impressive feature of the photograph above," declares LAWRENCE C. ROWAN, U.S. Geological Survey, "is the crater Kepler, 20 miles in diameter, near the center of it. The picture, taken by Lunar Orbiter III, shows an oblique view of the contact between the dark mare of Oceanus Procellarum in the south and the bright and rugged upland in the north.

"Kepler is generally believed to have been formed by meteorite impact. Chains of small craters alined

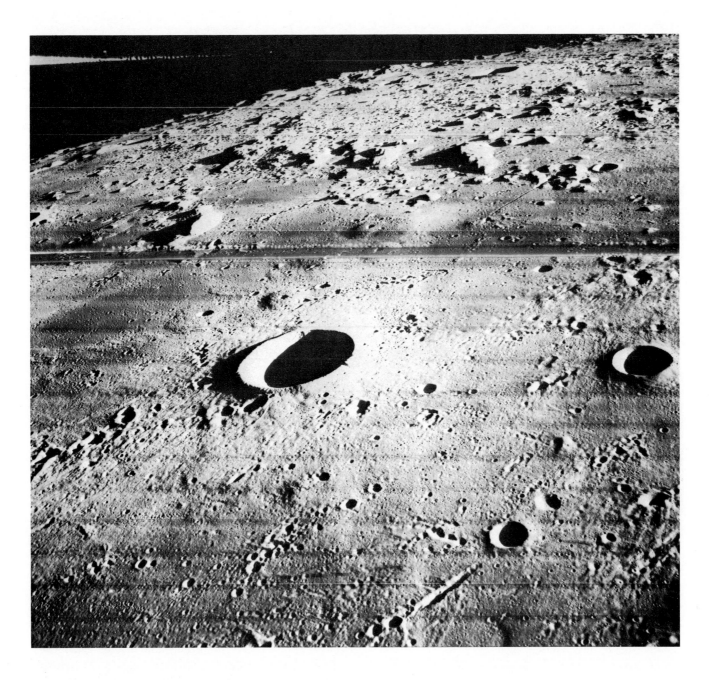

radially to it are superposed on both the mare and upland. They are most clearly defined in this photograph where they cross the mare south-southwest of Kepler. These chain craters appear as very bright rays on full-Moon photographs and were formed by impact of ejecta from Kepler. Superposition relations of this type show that the primary crater is younger than the subjacent material."

"The crater Hortensius [above]," says CHARLES J. DONLAN, then Associate Director, Langley Research Center, NASA, "is one of the smallest named craters, known chiefly for the domes north of it. This picture was taken by Lunar Orbiter III's 80-mm focal-length camera on February 20, 1967, from an altitude of 33.3 miles. The scene is remarkable not only for the fine detail of previously perceived features but for the abundant and clearly defined northeast–southeast tectonic structures."

"This view [above] encapsules the major events of lunar geologic history in the central part of the visible side of the Moon," said DON E. WILHELMS, U.S. Geological Survey. "It is one of many Lunar Orbiter III obliques whose objective was greater coverage than obtainable by vertical photography.

"The view looks northeastward at crater Murchison [55 kilometers in diameter], with part of crater Pallas at center left edge and part of crater Ukert [23 kilometers] at top, to right of center. Murchison, Pallas, and other craters, partly in view in the foreground, formed early, in pre-Imbrian time. These craters were then strongly affected in two ways by the formation, probably by impact, of the Imbrium Basin [outside picture to northwest]. They were cut by fractures radial to the basin; for example, the incomplete southeast rim of Murchison [at right edge] appears to be downdropped along faults. Second, the craters were partly covered by ejecta from the basin. The ejecta, which forms the rolling terrain in the upper fourth of the picture, has swamped the north rim of Murchison and spilled into the interior. Later events, in the time period called Imbrian, included the formation of the crater Ukert and the flooding of its floors, now cracked in places, by mare material and other volcanics."

100

"The moderate-resolution photograph on this page," explained Norman L. Crabill, of Langley Research Center, NASA, "was taken by Lunar Orbiter III on February 22, 1967, at an altitude of 35 miles, near the point 1° N, 58° W, in the southern part of Oceanus Procellarum, looking generally southwestward.

"This particular photograph was taken to determine whether the innermost crater, Damoiseau, of the large, double-walled crater at right center is a collapsed volcanic structure, as it appears to be from Earth-based telescopic observations. According to scientists of the U.S. Geological Survey, this oblique photo alone is not conclusive. However, a careful review of another Orbiter III photograph of the same area, a vertical view, indicates that the inner crater is indeed a collapsed volcanic crater and the outer wall the remnant of an older, impact crater.

"Another interesting aspect of this photograph is the contact between the marelike material in the foreground and the steep, semicircular cliff at its far edge. This headland has extremely steep sides, yet there is very little evidence of slumping at the base. This suggests that the mare material has flooded an old crater, its near rim already worn away by meteorite impact or submerged in general tilting."

"This oblique photograph [above] of the crater Theophilus was taken by Lunar Orbiter III on February 17, 1967," wrote E. C. DRALEY, Assistant Director for Flight Projects, Langley Research Center, NASA. "The spacecraft was looking south from a position near the Equator. The crater observed adjacent to Theophilus and almost at the horizon is Cyrillus.

"Theophilus is about 60 miles in diameter and approximately 4 miles deep. It has a very pronounced central peak, comparable in height to Mount Mitchell, the tallest mountain [6684 feet] on the east coast of the United States.

"This photograph will allow the visible characteristics of Theophilus to be compared with those of other large craters, such as Copernicus [pp. 116–117] and Tycho [pp. 120–121], that have been photographed by Lunar Orbiter spacecraft. This can now be done at a scale that was impossible prior to the Lunar Orbiter photography."

On the opposite page is a superb oblique photograph of the northern portion of Oceanus Procel-

larum. The Cavalerius Hills are in the foreground. The largest crater in the background is Galilei, about 10 miles across and over a mile deep. Its rims are 1000 feet above the surrounding terrain.

"Lunar Orbiter III made this picture from an altitude of 38 miles," said WILLIAM J. BOYER, Langley Research Center, NASA. "It was then 160 miles south of Galilei, just north of the Equator, and 1300 miles west of the center of the front face. The spacecraft was alined so that north in the picture is just left of top center.

"Contained in this frame of photography is the landing site of Russia's Luna IX [p. 50], which softlanded there February 3, 1966, the first spacecraft to do so. The best estimate of its location is near the central portion of the picture.

"I particularly remember," Boyer concluded, "that when this photograph was received from Lunar Orbiter III at the Control Center, in Pasadena, several of us spent many minutes scanning the negatives for some telltale shadow or reflection of Luna IX, but we were forced reluctantly to conclude that it was beyond the picture's resolution capabilities."

ORBITER III   103

ORBITER III

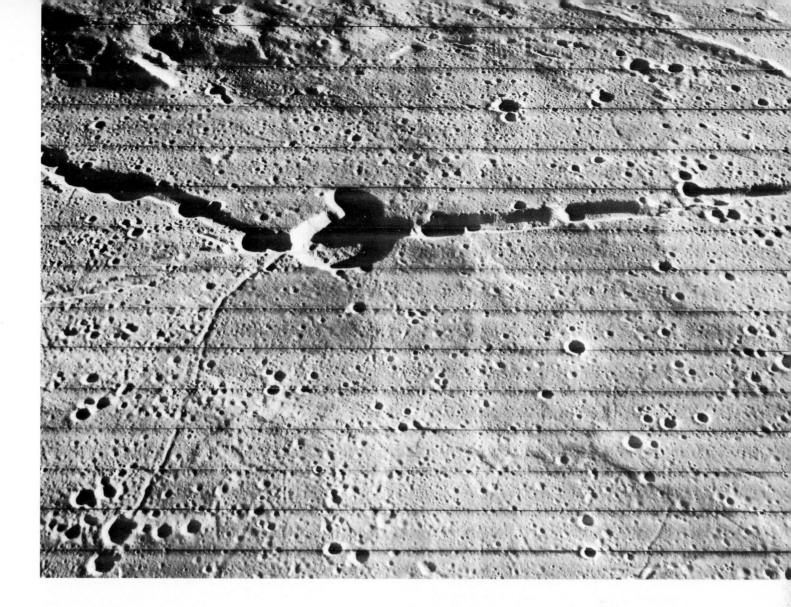

Stories in the Moon's Scars

"A boulder that rolled downhill and the track it made on the Moon are shown in the Lunar Orbiter photograph at left," wrote NEWELL J. TRASK, of the U.S. Geological Survey. "The track makes it evident that the surface material at this location [on the southeast interior wall of a mile-wide crater informally named 'Sabine EA,' at 2°48′ N, 24°54′ E] is soil-like, neither solid rock nor deep dust. North is at top, and the surface slopes gently from southeast to northwest. The track is 1200 feet long, 6 to 15 feet wide. Its freshness suggests that the boulder rolled relatively recently, probably dislodged by a small moonquake caused by a high-velocity impact nearby."

The dramatic Lunar Orbiter III photograph above is of the Hyginus Rille. HAROLD MASURSKY, Chief, Branch of Astrogeologic Studies, U.S. Geological Survey, explains that this rille is "located in the central part of the Moon. It's about 130 miles long, 2 to 4 miles wide, and about 1200 feet deep. It is a fault trough, or graben, with volcanic craters well developed along the western branch of the valley. The volcanic nature of the craters is indicated by the buildup of material around the vents; that is, the lips of the craters could not be made by collapse of material into the craters.

"The graben cuts rolling highland-basin fill material and mare material near the crater Hyginus, 8 miles in diameter, located at the bend in the valley. Exposed here are mare materials, upland-basin filling material, and the volcanic rocks near the craters. These vents may give suitable samples to study the composition of the Moon at great depths."

Some Studies in
Rocks and Rilles

Informative fine details of the Moon's surface not discernible from Earth are revealed in the Lunar Orbiter III photos explained on these pages.

"A fresh crater 500 feet in diameter is seen near the center of the high-resolution photograph [right] of the cratered mare surface of Oceanus Procellarum," wrote WILLIAM L. QUAIDE and VERNE R. OBERBECK, Ames Research Center, NASA. "It has a double-walled, concentric geometry peculiar to a size class of lunar-impact craters. Laboratory studies have shown that craters with this geometry are produced by impacts of projectiles against layered targets consisting of a loose surface layer lying on stronger subsurface rock. The outer part of the crater was formed in loose surface materials and the inner part in the subsurface rocks.

"Blocks of rocks ejected at low velocities from the central crater can be seen in the rays symmetrically disposed about the crater. Some of the ejected blocks have produced secondary craters upon impact against the loose surface materials. Many of the secondary craters have teardrop shapes. Some have sinuous furrows extending from the secondary crater in a direction generally outward from the parent concentric crater, terminating against the blocky secondary projectiles. The presence of these furrows further attests to the loose state of aggregation of the surface materials.

"The smaller, subdued-looking crater, 300 feet in diameter, below and to the right of the central one, has been severely eroded, but still retains evidence of its original concentric morphology. At an earlier time it may have had the same appearance as that of the fresh concentric crater.

"In fact," the commentators added, "craters of various geometric shapes and all degrees of freshness can be seen in this photograph, indicating that erosional processes must have continually modified the appearance of the lunar surface. This modification has been accompanied by repetitive bombardment of meteorites and cometary particles of all sizes.

Such repetitive impact has not just modified the appearance of the lunar surface, but, by crushing and grinding the originally hard surface rocks, it has produced the layer of loose material found there.

"Considerations of the size distribution of craters with morphology such as this has made it possible to measure the thickness of the surface layer at many places on the Moon, and thus determine the relative ages of formation of the hard rocks now covered by the loose mantle of debris."

"Characteristic features, heretofore not visible from Earth, of the floor of the northeastern part of crater Hevelius, near the far-west limb of the Moon, are impressively displayed in the photograph at top left, opposite page," points out VERL R. WILMARTH, Lunar and Planetary Programs, NASA.

"Hevelius is an old crater, about 60 miles in diameter, and similar to other large old craters. Many smaller, still younger craters are common in its floor and walls. The young craters—like the one, about 2 miles across, shown near the center of this photograph—are predominantly round, though a few are elongate. They have raised rims, but well-defined

throwout rays are not present. These features indicate that an extensive layer of weakly cohesive materials fills the crater and covers the crater walls.

"The most striking feature of this photograph," Wilmarth declares, "is the intersecting system of northwest- and northeast-trending rilles, trenches as wide as 3500 feet and several tens of feet deep. Two distinct ages of rilles are present. The oldest rille, not far from the left edge of the photograph and nearly parallel to it, is characterized by subdued walls. The other rilles are younger, have sharp walls, and are less modified by later surface processes.

"Rilles are interpreted as surface expressions of subsurface structural features of dynamic origin. This thesis suggests that two periods of diastrophism have occurred since Hevelius was formed. Similar structural features known on Earth are in some instances formed by upwarping of the surface rocks, by compaction, or by withdrawal of subsurface material."

"Part of an old lunar hill, one of several enclosing the basin of the ancient crater Flamsteed P., filled with mare material, is shown in the high-resolution photo at top right," wrote MARTIN J. SWETNICK, Lunar Orbiter Program scientist, NASA. "The hill, located at 42.8° W, 2.4° S, is the dominant feature in the picture, taken at a height of 32 miles, when the Sun was to the right at an elevation of 19° above the local horizon. Although the area covered is 3.3 by 2.7 miles, surface features as small as 4 feet in diameter can be readily resolved. North is toward the upper-left corner of the photograph.

"The crest of the hill is 1000 feet above the cratered floor of the basin," Swetnick continued. "The western, or shadowed, side of the hill has a slope of about 18°. The slope of the eastern side is less than 10°. The features of the western slope have scientific significance in that they provide evidence for the occurrence of mass wasting. A terrace, or talus slope, consisting of fragmented material can readily be seen along the base of the western slope. The terrace was most likely formed by a landslide down that slope. Ledge rock exposed by this landslide can be seen near the crest of the hill. The texture of the terrace material is distinctly finer than that of the material higher on the hill slope. The density of craters on the terrace is much less than on the basin's floor."

ORBITER IV

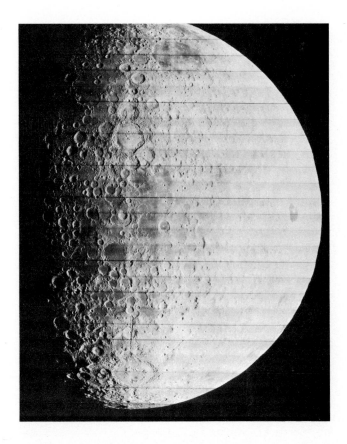 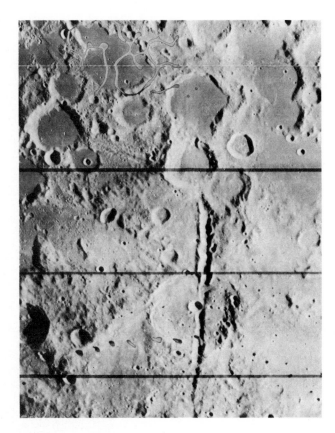

Orbiter IV Maps the Moon's Near Side

"The prime objective of Lunar Orbiter IV," wrote JAMES S. MARTIN, Deputy Lunar Orbiter Project Manager, Langley Research Center, NASA, "was to photograph the entire front side of the Moon at a resolution considerably better than is possible from Earth. The mosaic on the opposite page is the result. The pictures, taken May 11–25, 1967, span an area from the North Pole (at top) to the South Pole, and from the eastern limb (at right) to the western limb.

"The complete mosaic is approximately 40 by 45 feet. It was laid on the floor, and observers were allowed to stand on it or crawl over it in their stocking feet. Some astronomers chose the latter course, carrying magnifying glasses. The mosaic was a primary source in selecting scientific sites for Lunar Orbiter V to photograph at higher resolutions."

"The first swath of Lunar Orbiter IV's nearly polar orbit of the Moon on its mapping mission disclosed some hitherto unseen areas beyond the eastern limb," commented LEON KOSOFSKY, Lunar and Planetary Programs, NASA. "One of the most striking features disclosed was the large trough visible in both photographs above.

"This great gash [at top left of the biggest crater in the wide-angle photograph; enlarged at upper right] is about 150 miles long and averages 5 miles in width. It is the most prominent of 3 valleys radiating from that large, double-rimmed crater. Presumably the valleys were formed by the impact that created the crater. There are radial valleys near other very large craters. As the enlargement shows, the edges of this trough are raised and scalloped, giving it the appearance of a chain of closely spaced craters."

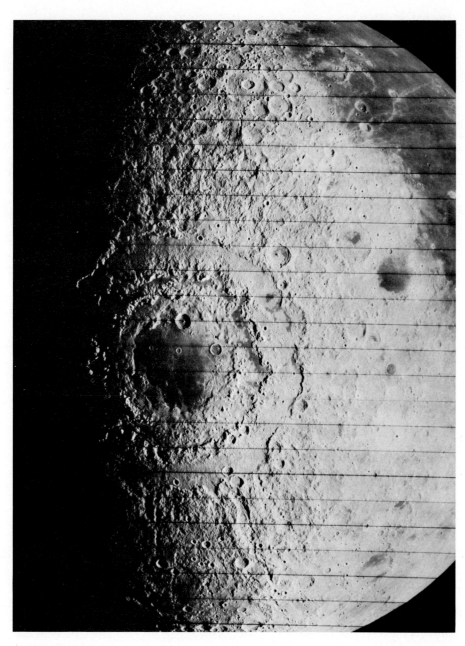

A Crater
as Big as
New York State

"Lunar Orbiter IV was about 1700 miles above the Moon on May 25, 1967, when its wide-angle and telephoto cameras simultaneously took the two dramatic photographs paired on these pages," said A. THOMAS YOUNG, Lunar Orbiter Project Office, Langley Research Center, NASA. "The wide-angle photograph [above] shows the large, circular Orientale Basin, appearing as a gigantic bull's-eye on the western limb of the Moon. The Cordillera Mountains, rising some 20 000 feet above the adjacent

surface, form the outer ring, which is approximately 600 miles in diameter. The State of New York would easily fit within it.

"The telephoto view on the opposite page magnifies in remarkable detail the portion of the wide-angle photograph showing the area just to the right of center of the Orientale Basin.

"Large, circular basins such as Imbrium, Crisium, and Orientale are major features on the Moon, and therefore of great importance to lunar science. The

110

sharpness of the features portrayed in these photographs has led many lunar scientists to interpret Orientale to be one of the youngest of the large basins. Orientale's appearance is considered to be typical of how the older basins, such as Imbrium, looked earlier in their long history."

The Orientale Basin is centered at 89° W, 15° S, on the extreme western edge of the Moon's visible side. It is interpreted as having been formed by the impact of a giant meteorite or comet nucleus, and as being only partly filled by the younger, dark volcanic material.

Its outer rim is covered by ejecta hurled from the basin. Within Orientale's outer ring, the Rook Mountains, seen in the upper-right portion of the telephoto view, form another circular scarp, about 400 miles in diameter. At the center of the Orientale Basin is the dark Mare Orientale, which is about 186 miles in diameter.

Moon "Dunes" Against a Crater Wall

"The crater Riccioli, occupying the right center of the high-resolution Lunar Orbiter IV photograph on the opposite page, is about 130 kilometers in diameter and lies about 750 kilometers to the northeast of the center of the Orientale Basin," explained J. F. McCAULEY, of the U.S. Geological Survey.

"Pre-Orbiter telescopic work by Hartman and Kuiper, of the Lunar and Planetary Laboratory, University of Arizona, and by me suggested that Orientale was the youngest of the large lunar basins. Orbiter IV convincingly proved this hypothesis.

"The well-preserved, grooved-to-braided texture of the 'blanket' that surrounds Orientale can be seen in the accompanying photograph. The fact that it mantles pre-Orientale craters, such as Riccioli, is also evident. The blanket is clearly draped over the crater wall nearest Orientale, but is banked up into crude dunes on the far wall [upper-right side of Riccioli]. The material of the blanket apparently 'flowed' over the rim nearest Orientale and out over the crater floor, but, upon reaching the far wall, lacked sufficient energy to move in an upslope direction.

"Similar 'dunelike' deposits are found at the base of topographic obstacles at comparable distances from Orientale around the entire basin. Near the Cordillera Mountains, closer to the center of Orientale, preblanket craters are completely mantled, and the 'banking-up' effect cannot be seen. These features are interpreted to be the product of surface flow in a base-surge cloud of the type known to occur in nuclear and high-explosive cratering events.

"The new data contained in the Orbiter IV photographs of the Orientale Basin," McCauley concluded, "strengthen the hypothesis of an impact origin for large lunar basins."

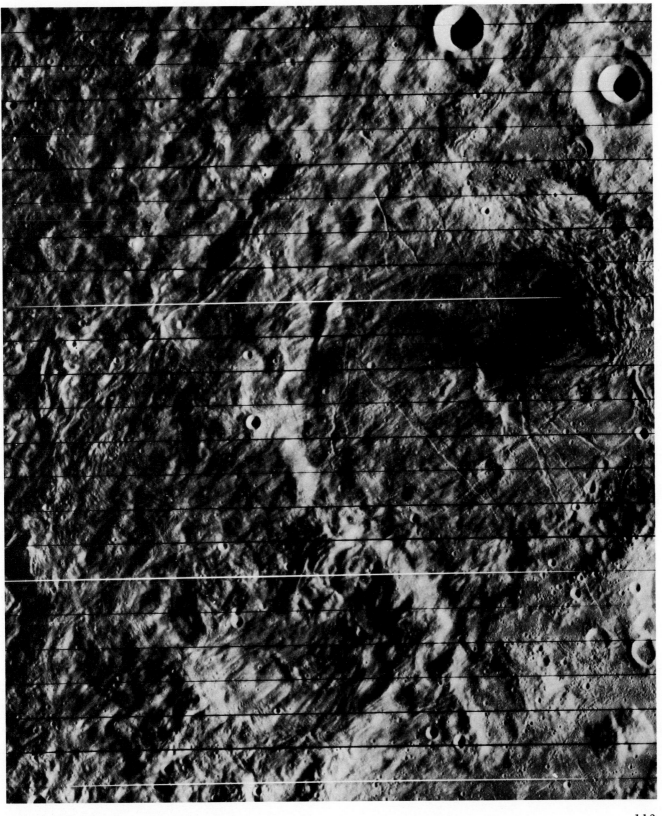

Orbiter V Photographs the Nearly Full Earth

"This photograph of the Earth [opposite page] was made by Orbiter V as it circled the Moon in a polar orbit 214 860 miles away," commented Lee R. Scherer, Lunar Orbiter Program Manager, NASA. "It does not quite include the full sunlit surface. The picture is centered over India, while the Sun was directly over Saudi Arabia.

"The clear outline of the entire east coast of Africa and the Cape of Good Hope can be seen. Prominent geographical features such as Italy, Greece, Turkey, the Black Sea, the Suez Canal area, the Red Sea, and the Persian Gulf are discernible.

"This was one of the secondary photographs of Orbiter V's mission, but it attracted high public interest around the world for a variety of reasons.

It was the first picture of the nearly full Earth as seen from the Moon distance.

"Wags examining this photograph have stated that life obviously could not exist on such a shrouded planet. To most, this was humorous, but thoughtful scientists noted the distinct resemblance to photographs of Venus, and interest in the exploration of that planet was heightened.

"To the philosopher, this picture is close to man's soul, since it allows him to look back upon his own world, and thus fosters the feeling of man's emancipation from the bounds of Earth. Perhaps to most of us the interest is due to human vanity, which dictates that self-portraits are always the best portraits."

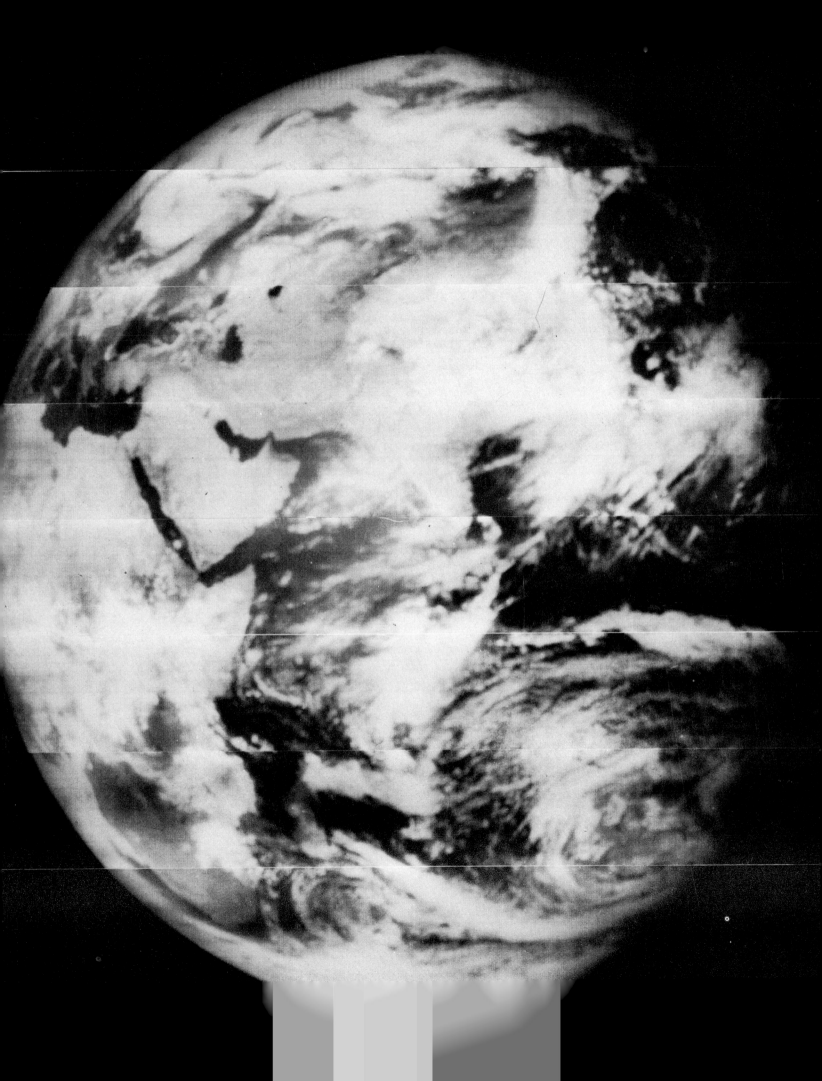

A Closer Look at Copernicus

"The medium-resolution [above] and high-resolution photographs of the crater Copernicus shown here," said Terry W. Offield, of the U.S. Geological Survey, "were taken by Lunar Orbiter V from an altitude of 63 miles above the Moon's surface in August 1967." The area within the rectangle, shown enlarged on the opposite page, is 12½ miles wide and 17½ miles long.

"Copernicus [see also pp. 88–89] is relatively young and fresh. The medium-resolution photo-

graph shows a hummocky crater rim, numerous large slump blocks on the crater wall, and a complex of central peaks. Sets of parallel fractures, alined with the lunar structure grid, formed after the crater wall took its present form, but before the smoothest floor materials were emplaced. The smooth floor materials show a swirling pattern of cracks like those seen on terrestrial lava flows. These materials are associated with numerous hills that have summit craters and are probably small volcanoes. Several low places on the rim and wall are partly filled by what appears to be ponded volcanic material, or possibly fluidized impact debris.

"The high-resolution photo," Offield concluded, "shows that flows apparently spilled through breaks in the pond walls and coursed down valleys to the crater floor, where they spread out as deltas or lobes. Ridges on the flows are natural levees, formed as material solidified faster along the channel sides than it did in the center."

Lunar Tales
in Tracks
and Troughs

"The tracks of numerous rolling boulders, some of them more than 100 feet across, are visible in Lunar Orbiter V's high-resolution photograph of the crater Vitello, on the southern edge of the Moon's Sea of Moisture," remarked KENNETH L. WADLIN, of the Lunar and Plantetary Programs, NASA. "The picture was taken on August 17, 1967, from an altitude of about 100 miles.

"Two such tracks can be seen in the enlargement at right. The complex pattern of the 900-foot track left by the large boulder indicates that this boulder, which is about 75 feet across, is quite angular. The buffeting that it experienced in rolling did not cause it to break up or become rounded. This points both to the integrity of the material of which it is composed and to the deformable nature of the surrounding surface material.

"The more uniform track of a smaller, more rounded boulder can be seen to the right of the large one. This smaller boulder is over 15 feet across, and traveled 1200 feet before it came to rest.

"The boulders may have been set rolling by moonquakes or meteor impacts."

Orbiter V took more revealing photographs of the Hyginus Rille [top of opposite page] than Orbiter III did [see p. 104].

"The moderate-resolution photograph at top left shows clearly the two elements of the rille system," explained HAROLD MASURSKY, Chief, Branch of Astrogeologic Studies, U.S. Geological Survey. "First is the trough bounded by marginal faults. The graben, or downdropped block, is nearly a thousand feet below the surrounding plain. Disposed along the west branch of the fault valley are a series of volcanic vents, like beads on a string. The largest vent, the crater Hyginus, located at the bend in the valley, is 8 miles in diameter.

"Details of the fault valley and volcanic vent are more clearly visible in the high-resolution photograph at top right. In particular, abundant outcrops and boulders occur on the lip and sides of the vent. These blocks are the critical feature for, in terrestrial volcanic vents that greatly resemble these lunar ones, blocks occur that are brought up from great depths. In the Colorado Plateau area of the United States, blocks around such vents as these have been brought up from depths of 20 miles.

"It should be possible to make a manned landing in the relatively smooth topography adjacent to this valley," Masursky concludes. "From this landing site, it should be possible to collect samples of typical upland materials from the walls of the trough, as well as samples from the vent that are brought up from deep within the Moon."

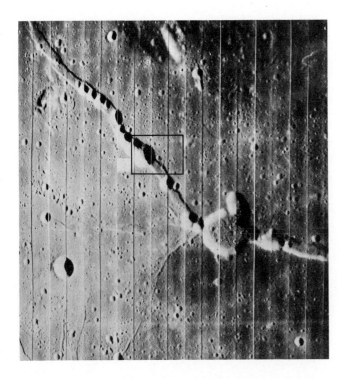

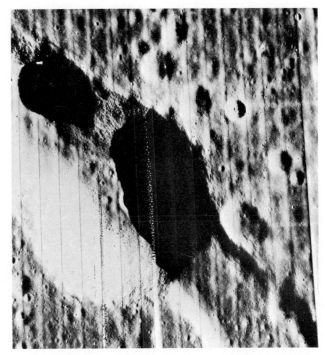

Below are moderate- and high-resolution photos by Lunar Orbiter V of sinuous rilles in the Marius Hills. ROBERT P. BRYSON, Lunar and Planetary Programs, NASA, gave this description: ". . . they are long, uniformly narrow rifts or valleys that, like river channels, pass across or through a 'wrinkle ridge' and between domes, or low hills. Unlike river channels, however, they appear to head in a basin or crater and decrease gradually in width and depth and disappear. Very likely they were formed by some type of freely flowing material—possibly water [as Dr. Harold C. Urey has suggested], although, under the existing condition of low gravity and atmospheric pressure, as well as broad temperature extremes, water could have remained at the surface for no more than very brief periods."

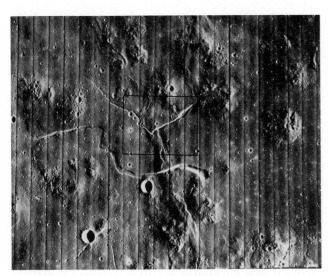

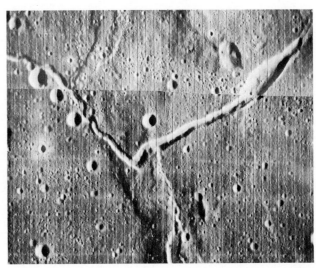

The Intricate Structure of Tycho

"The youngest large impact crater on the Moon is Tycho, which is 53 miles in diameter and nearly 3 miles deep," commented HOWARD A. POHN, of the U.S. Geological Survey. Surveyor VII [see p. 78] landed about 18 miles beyond its north rim, as shown in the medium-resolution view of Tycho above,

taken by Lunar Orbiter V. [The white circle marks its landing area. The area within the rectangle, enlarged on the opposite page, is 27 by 35 miles.]

"This photograph shows the entire crater, including ejected crater materials, which form pronounced radial ridges and grooves. These ejected materials

are gradational from large, hummocky, angular blocks, nearest the crest of the crater rim, to fine, subdued materials, near the top of the photograph.

"The crater interior is characterized by a pronounced central peak, extremely hummocky floor, and steep, angular blocks in the walls. Many depressions along the eastern margin of the crater interior exhibit ponds of smooth materials that are thought to be the result of postcrater volcanism.

"The high-resolution photograph above shows the intricate structures that appear inside the crater, with the most interesting feature being the crater floor. The subparallel ridges and grooves, as well as the many cratered prominences, are reminiscent of the Bonita Lava Flow, in northern Arizona.

"The floor of Tycho is demonstrably younger than the crater rim, and was probably formed much later in lunar history, as a result of tapping deep-seated volcanic sources by fracturing that was associated with the Tycho impact."

Penetrating Views
of Aristarchus

"The youthful impact crater Aristarchus, viewed above in medium-resolution and on the opposite page in high-resolution photography by Lunar Orbiter V, is 23 miles in diameter and 10 000 feet deep," wrote DONALD U. WISE, associate professor of geology, Franklin and Marshall College. "It illustrates the interplay of geologic processes that complicate lunar land forms. Three major topographic zones are characterized by differing age and dominant geologic process. [The area within the rectangle, enlarged on the opposite page, is 11 by 14 miles.]

"Oldest of the zones is the debris apron, spreading outward for a crater diameter beyond the upturned rim. The second zone, a ring of slumped blocks, resulted from gravity collapse of the initial steep walls of the crater to produce the present, relatively stable 40 percent slopes. Many of these collapses utilized a preexisting northeast- and northwest-trending regional fracture system to produce an intersecting array of elongate slump blocks and an eyelike north-south elongation of the crater [visible above].

"The third and youngest zone, strikingly similar in appearance to floors of Hawaiian volcanoes, is relatively flat, strongly crevassed, and marked by many dome-shaped hills, 50 to 100 feet high. Absence of major disruption of the flat floor by slump movement indicates that volcanic filling of the floor is younger than most of the wall collapse. Many late-stage modifications of the slumped-wall zone may have been coincident in time with the volcanic activity. Periodic sightings of red glows in the Aristarchus region suggest that this phase of crater modification may still be in progress."

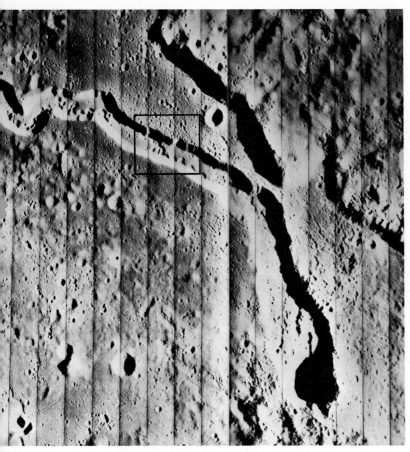

Schröter's Valley and Some Sinuous Rilles

"A lunar valley named after the renowned German selenographer Johann Hieronymus Schröter (1745–1816) is depicted in the excellent photographs above," said DENNIS B. JAMES, of Bellcomm, Inc. "The medium-resolution [left] and high-resolution photographs of it were taken simultaneously by Lunar Orbiter V on August 18, 1967, from an altitude of about 140 kilometers. The distance between vertical lines in the medium-resolution photograph above is 4300 meters.

"Schröter's Valley terminates in what lunar scientists have been inspired to call (for obvious reasons) the Cobra Head [bottom right in the photo], which is interpreted as its source. The valley's walls rise more than 1300 meters above its floor. It is near the craters Herodotus and Aristarchus, but closer to the former.

"In the cropped telephoto view at right above, the salient feature displayed is the meandering rille within Schröter's Valley," James continued. "The rille also emerges from the Cobra Head. As it me-

anders, more so than the valley itself, the rille moves from one side of the valley to the other, at one point forming a cutoff channel. These characteristics resemble those of meandering canyons, valleys, and streams on Earth. Of special interest also are the blocks of rock that separate from the valley's walls and leave behind them tracks of their downslope movement.

"Scientists believe that the valley and its meandering rille were formed by erosion of small particles carried by a flow of either a liquid or a gas.

"The distance between vertical lines in the high-resolution photo is 540 meters."

"Sinuous rilles are among the most unusual features of the lunar surface," declared MARTIN W. MOLLOY, program scientist, Apollo Lunar Exploration Office, NASA. "Perhaps billowing clouds from the fluid that carved meandering channels like those in the Harbinger Mountains, shown in the Lunar Orbiter V medium- and high-resolution photos at

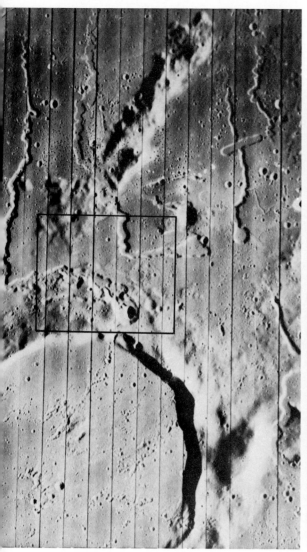

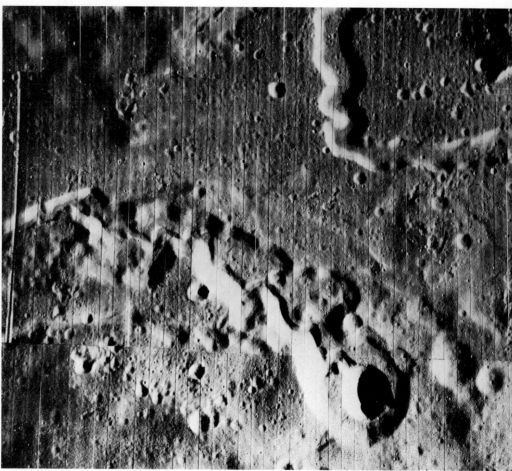

the top of this page, boiled off in the ultra-high vacuum of the Moon and distributed fine particles uniformly over adjacent areas.

"The sinuous rilles along the right edge of the left-hand photograph show evidence of subsequent channel filling by mare material. The rille in the top center of the picture has cut through an older line of hills in Oceanus Procellarum. The large crater at the bottom is Prinz. Its floor is spattered by secondary ejecta from Aristarchus, just out of sight at lower left. Its rim is partly submerged in the material that has flooded Oceanus Procellarum.

"Flowing downhill [toward the top of the photo] from the rim of Prinz and out onto the mare surface is a rille with a narrow, new channel incised in the side of an older, broader channel.

"This same rille is seen in detail in the high-resolution photograph at right," Molloy added. "A small, recent crater located within the head of the older rille is the source of the new channel. The crater is so fresh that large blocks can be seen in its walls."

The Moon's
Pale Self and Distant Views

Though Surveyor I's television camera produced black-and-white images, one way existed by which it could transmit color. This was to shoot three separate pictures of the same scene through different-colored filters, send them to Earth, and there reconstitute the picture through similar filters. One result of this complicated process was the photo above, which indicates that the lunar surface appears to be almost colorless.

In explanation of the picture, CLARK GOODMAN, of the University of Houston, said: "Since solar light is essentially white—composed of equal amounts of all wavelengths over the spectral range of the photographic film—the light reflected from the Moon is solar light minus any wavelengths specifically ab-

sorbed by the lunar grains. The fact that the reflected light is nearly colorless (nearly white) means that the lunar skin does not selectively absorb any particular wavelengths.

"Hence," said Goodman, "we conclude that the lunar material is not composed of a particular single-colored mineral but instead is either a mixture of many different colored minerals or is a mixture of colorless minerals."

Another color reconstitution, on the opposite page, is a closeup of Surveyor III's footpad 2.

Remarking on it, J. J. RENNILSON, of the Jet Propulsion Laboratory, said: "Visible items of importance, from upper left to lower right, are: a small

portion of lunar soil, placed on the footpad by the surface-sampling instrument; the attitude-control jet, with its gold-tipped nozzle contrasting with the gray of the lunar soil; and, lastly, a photometric target, consisting of several shades of gray as well as three colors.

"One of the major scientific achievements of the color analysis of such pictures as this one was the discovery that very little color differences were detectable between the immediate surroundings of Surveyor I's landing site and those of Surveyor III's landing site. The color of the disturbed soil remained the same as the color of the undisturbed surface."

"The color-reconstituted photograph at top right," said J. J. RENNILSON, of the Jet Propulsion Laboratory, "is the first one in which man has been able to observe an eclipse of the Sun by his own planet. Surveyor III took the view from the Moon with the wide-angle mode of its TV camera.

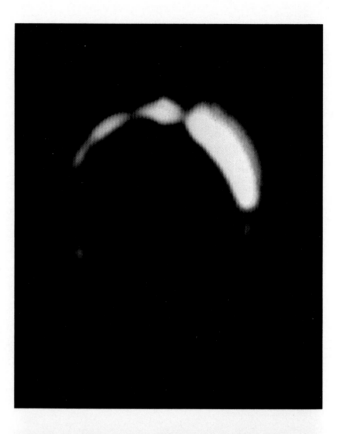

"Most prominent in the picture is the white cap of light caused by the bending of the Sun's light as it passed through the Earth's atmosphere. The cap is much brighter than the rest because of the Sun's proximity to that limb, causing a greater proportion of sunlight to be refracted. The beaded appearance around the remaining portion of the Earth's atmosphere is due largely to the interruption of the band of light by overcast areas. A small portion at the right of the solar-eclipse photograph was obscured by an edge of the camera's mirror.

"Blue light from the Sun is scattered out of the beam during passage through the Earth's atmosphere, leaving mostly the green, yellow, and red portions of the spectrum. A careful study of this and other lunar photographs of the solar eclipse will enable scientists to understand better the optical properties of our atmosphere."

"The historic color photograph of the crescent Earth [right]," commented ROBERT F. GARBARINI, former Deputy Associate Administrator (Engineering), Office of Space Science and Applications, "was taken from the Moon's surface by Surveyor III on April 30, 1967. Only because of the tilt of the spacecraft and the favorable libration of the Moon was it possible to catch the Earth in the camera's wide-angle field of view."

On a later Surveyor mission, it was possible to command the camera to take narrow-angle photographs of the Earth having much higher resolution.

Earth's First Close Views of Mars

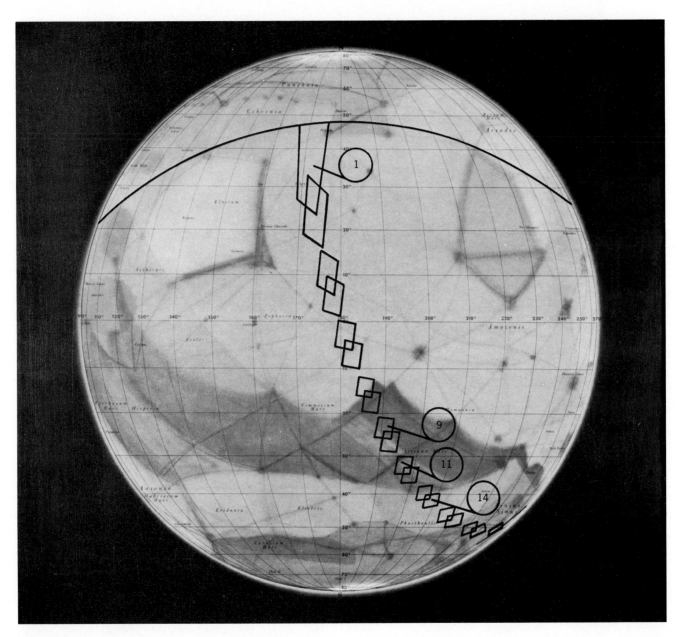

The photographs on the four following pages were taken by a television camera that was nearly 150 million miles away, farther from Earth than any camera had ever traveled before. It was riding on Mariner IV, which, after an interplanetary journey of 7½ months, flew within 9847 kilometers of Mars on July 15, 1965.

The course of the spacecraft camera's picture coverage of Mars and the locations of the pictures on pages 130–133 are shown on this map.

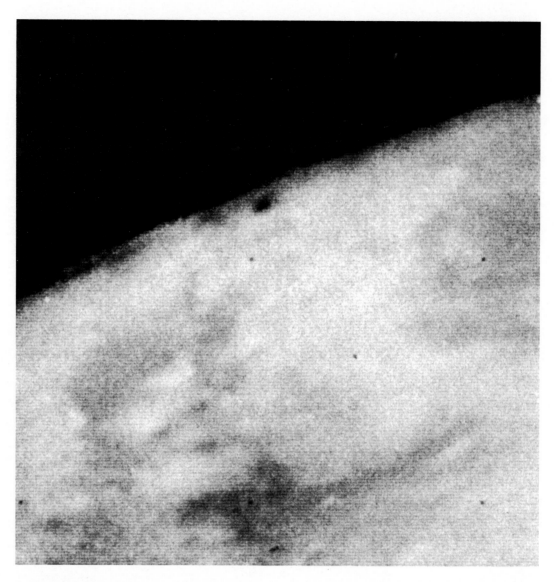

Mariner IV's camera, its shutter automatically operating every 48 seconds as red and green filters were alternated before its lens, took a total of 21 complete pictures and a fraction of another.

These pictures provided man with his first chance since the invention of the telescope to view the surface of Mars without the hindrance of "straining to see through the Earth's atmosphere like a driver peering through a rain-spattered windshield," as Dr. Robert Jastrow has expressed it.

"At 00:18:33.1 GMT on July 15," said DAN SCHNEIDERMAN, Mariner 1967 Project Manager, Jet Propulsion Laboratory, "the first closeup black-and-white photograph of Mars [above] was taken from the spacecraft through a red filter. The image center is approximately at 33° N, 171.6° E, and about 17 000 kilometers from Mariner IV. The spacecraft velocity relative to Mars was 4.913 km/sec.

"Surface features are not readily distinguished because of what appears to be a haze or cloud layer, which is enhanced by the oblique view. There is an uncertainty as to the condition of the lens at this time, and some glare may be present. The upper right-hand region corresponds with the area known as Phlegra. Because of the cloud layer, it cannot be determined whether the light and dark areas are due to changes in aspect of the surface, a series of breaks in the clouds, or a combination of these effects.

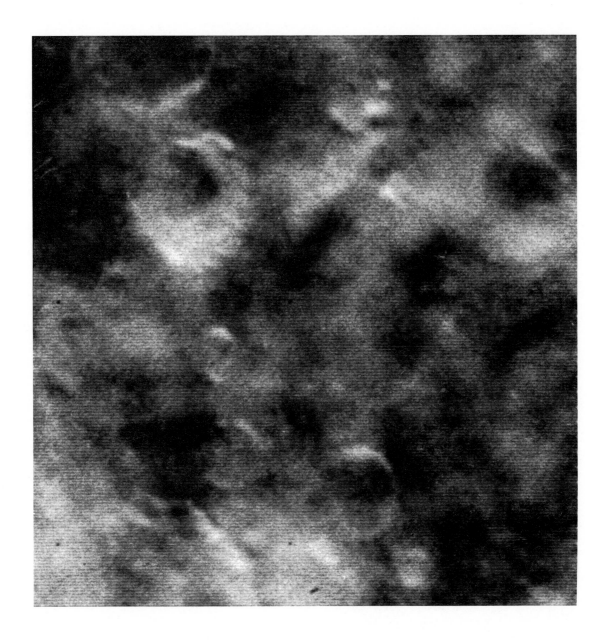

"The noteworthy aspect of this Mariner IV photograph [frame No. 14] of the Martian surface," declared ROBERT P. SHARP, of the California Institute of Technology, "is not that it shows more, larger, or better craters, but that several craters are outlined in white. This has been interpreted as a crusting of frost. Irregular white spots in the north-central part of the photo may be frosted areas in an irregular highland.

"The location is 41° S latitude within the light area Phaethontis. The area shown above is 140 by 170 miles. The time is late afternoon in midwinter, with the Sun to the north and 30° above the horizon.

"During winter at this latitude, frost could form each night over much of the landscape and might survive through the day in favorably shadowed and properly located situations. The edge of the south polar cap is only 10° farther south at this season. More speculative is the nature of the frost. The most acceptable current model of the Martian atmosphere suggests that it may be solid CO_2 rather than water frost."

MAN'S VENTURES

INTO SPACE

MAN'S VENTURES INTO SPACE

MAN HAS BEEN EXPLORING space since 1961. By 1968, 30 men and one woman had accumulated more than 100 days of space-flight experience. It is now certain that men can live in the weightless environment of space for periods of up to 2 weeks and return to Earth without apparent ill effects. In addition, men have learned—

- To control and navigate spacecraft.
- To use tools, cameras, and scientific instruments in flight.
- To operate outside a spacecraft protected only by a spacesuit.
- To carry out rendezvous and docking with another spacecraft launched separately.
- To guide a returning spacecraft through the atmosphere to a selected landing area.

Thus far men's activities in space have been limited to the region immediately surrounding their native planet. The deepest venture into space during NASA's first decade was that of Astronauts Charles Conrad and Richard Gordon, whose Gemini XI spacecraft, docked with an Agena propulsion system, climbed to an altitude of 853 miles over Australia in September 1966. But as the Apollo space vehicle and its supporting ground equipment approach readiness, men are poised for a leap across the void in the mission to the Moon, a quarter-million miles away.

The first man to depart from Earth was a Soviet cosmonaut, Yuri Gagarin. His Vostok I carried him around the world on April 12, 1961. Later that year another Soviet cosmonaut, Gherman Titov, spent a day in orbit before returning safely, and two American astronauts, Alan Shepard and Virgil Grissom, made suborbital flights of 300 miles to test the Mercury spacecraft.

Orbital flights by American astronauts began in 1962. John Glenn and Scott Carpenter made three revolutions each, and Walter Schirra extended this to six revolutions for a total of 9 hours in flight.

That same year Soviet Cosmonauts Andrian Nikolayev and Pavel Popovich carried out the first dual space mission involving two vehicles launched separately. Nikolayev went into orbit first and Popovich followed a day later. Both stayed aloft for an additional 3 days, in orbits that brought their craft within 4.3 miles of one another. In another dual mission the next year, Cosmonaut Valery Bykovsky was in orbit 5 days and a Soviet woman, Valentina Tereshkova, was aloft 3 days; they came within 3 miles of each other in space.

The United States concluded its Mercury program in 1963 with a flight by Gordon Cooper that lasted 34 hours. The first spacecraft to carry more than one passenger was the Voskhod I in 1964. In its crew were the late Vladimir Komarov, a pilot; Konstantin Feoktistov, a scientist; and Boris Yegorov, a physician. The photographs on the next four pages were taken from American single-passenger spacecraft. The others in this section were taken from the two-man Gemini; the United States first launched its Gemini for testing without a crew in 1964.

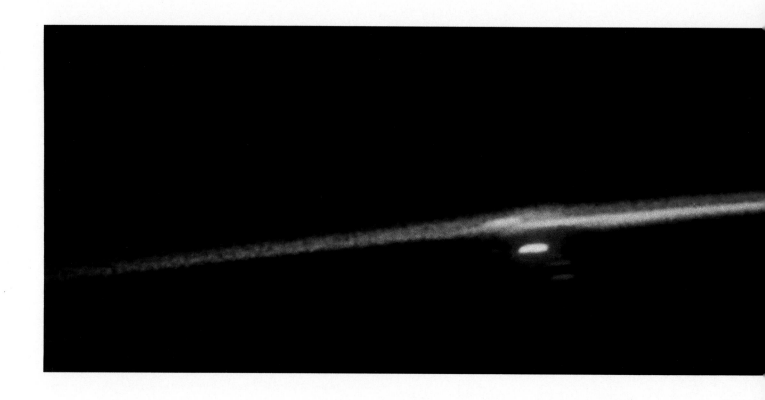

From Cameras Held by Men

"Sky is the part of creation," John Ruskin wrote long before men soared into it, "in which Nature has done more for the sake of pleasing man, more for the sole purpose of talking to him and teaching him, than in any other of her works. . . ."

American astronauts groped for words to describe their feelings when they looked homeward from the sky, and hastened to record on film what they saw.

Astronaut SCOTT CARPENTER said of the sunset above: "The beauty of this panorama is overwhelming. The brilliance of the colors and the sequence in which they appear defy description, and I was anxious to record it photographically so others would understand. Even at that time, however, I was consoled by the thought that, in contrast to the ever-changing aspect of sunrises and sunsets as seen from the surface of the Earth, those viewed from space were destined to be forever the same."

Astronaut JOHN GLENN wrote of the photo at right: "This picture of the Atlas Mountain range in Morocco was taken during the first orbit of the FRIENDSHIP 7 flight on February 20, 1962. Since this was the first U.S. manned orbital flight, this was the first landmass picture taken following launch approximately 23 minutes earlier. On this flight I used a 35-mm hand-held camera. Probably the most significant thing about the photograph is that it pointed out to all of us the value pictures from space would have for mapping, weather analysis, etc., in work that has since been refined to a high degree with other equipment on other flights and other projects."

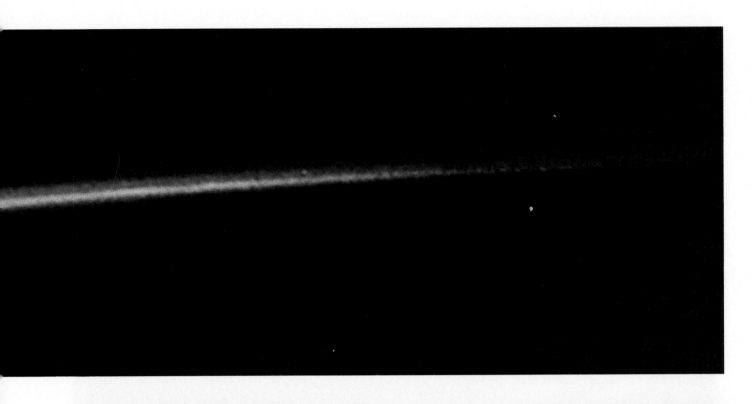

AURORA 7

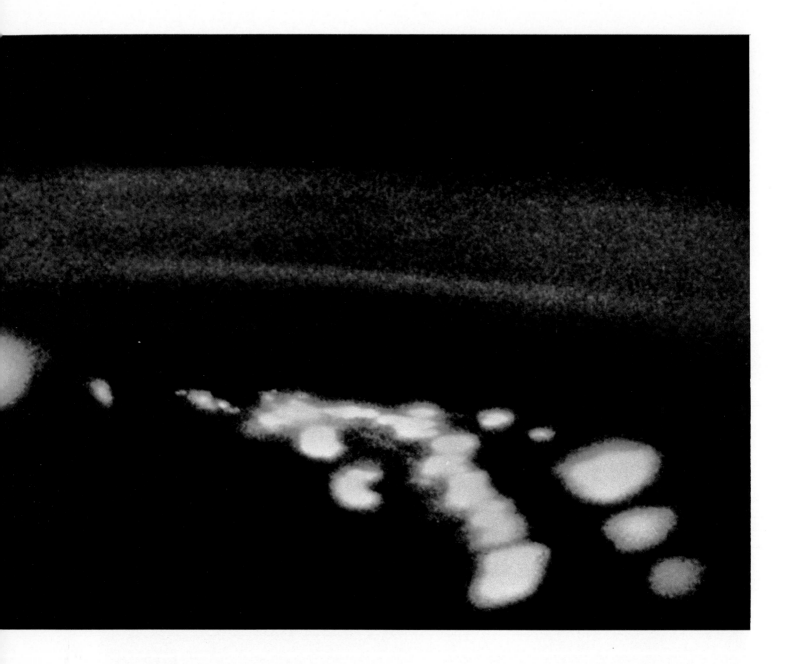

John E. Naugle, Associate Administrator for Space Science and Applications, NASA, recalls: "This is the first photograph of the night airglow taken from above. It was taken by Gordon Cooper during his orbital flight of May 15, 1963, for an experiment conducted by Edward P. Ney, W. F. Hutch, and F. C. Gillett of the University of Minnesota. The photograph is a 2-minute exposure taken early in the satellite night as the spacecraft faced backward over the eastern coast of Australia. The flashes in the foreground are individual lightning discharges from four active regions in a storm. The successive flashes recede into the distance as the spacecraft moves away from the storm. The airglow layer, very much smeared by spacecraft motion, is the greenish band which begins in the photo about an inch above the horizon as delineated by the lightning flashes in the distance. This and other photographs show an airglow layer 24 kilometers thick and 77 to 110 kilometers above the horizon."

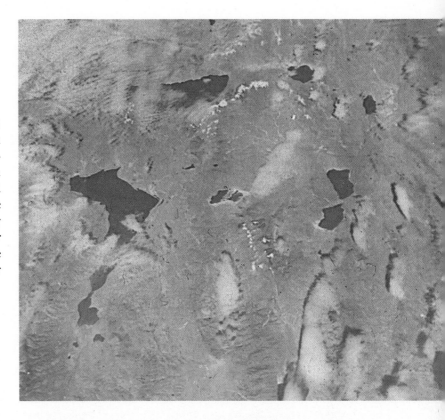

Astronaut GORDON COOPER wrote of this picture: "The Tibetan lake country was very distinctive in that our orbits were over primarily lower altitude terrain of Arabia and India—then rather abruptly you came upon this higher terrain with sandy-colored soil, interspersed with numerous lakes. You can see patches of snow scattered about the ridges and many of the lakes are partially choked with ice. The air is very clear in this area. This is one area where brown and green colors appear as they are, rather than with a bluish tint from haze and humidity."

ROBERT R. GILRUTH, Director of the Manned Spacecraft Center, NASA, points out: "This photo of the Himalayan region is almost vertical from an altitude of approximately 90 nautical miles. The China-Nepal border zigzags across the lower half of the picture. The largest single feature, approximately in the left center, is the completely snow-covered Gurla Mandhata Massif, with a peak elevation of 25 335 feet. The high peaks in Nepal are cloud covered, but the clear view of the Tibetan region shows much hitherto unknown detail of this remote and poorly explored area of Asia."

FAITH 7

A New Art Germinates

The first two Gemini flights demonstrated the structural integrity of the U.S. first two-passenger spacecraft and the performance of the launch vehicle and spacecraft systems. The third was a three-orbit manned flight, and the fourth a 4-day manned flight, on both of which the astronauts photographed the Earth.

The next seven pictures are examples of the results. Astronauts James A. McDivitt and the late Edward White took them in June 1965 during their 4 days in space. The camera used was a modified Hasselblad, Model 500 C, with a Zeiss planar lens of 80-mm focal length, and the film was Ektachrome MC (S.O.–217). The highest altitude reached was 175 miles, and the lowest point while in orbit was 100 miles.

The 219 color photographs brought back by the men aboard Gemini IV demonstrated the potentialities of viewing the ocean's surface and floor, the shorelines of continents and islands, and the great variations in terrain from high altitudes.

The photograph on the next page, for instance, was of especial interest because of a cautionary statement in the U.S. Naval Oceanographic Chart HO–5394, "Crooked Island Passage," dated November 1965. It read: "The charted position, size, shape, and ori-entation of the islands in the Bahama Islands are unreliable." When the outlines of the charted islands and those in photographs of the Bahama Islands taken from space were compared, the validity of this warning was readily apparent. These islands are close to the United States and men have sailed through the passages between them for hundreds of years, yet never could see the contours before in the way they were visible in photographs taken from spacecraft.

On this same flight, two famous river mouths, those of the Nile and the Colorado Rivers, were photographed as entities for scientists examining interactions between such features of the Earth and men's activities. These unprecedented views are reproduced on pages 144 and 145. The picture of the Colorado River's mouth was one of a series of overlapping photos of North America which, when viewed together, tell scientists more than can be ascertained from a single photograph.

The other Gemini IV pictures shown here include desert areas of both the Old World and the New World, an enormous structure in Mauritania that has long puzzled geologists, and finally the Florida Keys, an area with which many Americans have long been familiar.

142

"Acklins and Crooked Islands in the Bahamas," says ROBERT E. STEVENSON of the Bureau of Commercial Fisheries, Department of the Interior, "are typical of shoals and islands that rise abruptly from depths of 2000 meters in these waters. Sunlight reflecting from brilliantly white calcareous sand at depths less than 5 meters produces the vivid light-blue color around the islands. The wave pattern and current eddies in the reflection from the Sun are formed around nearby banks, with narrow, bright, irregular sea slicks indicating the water motion. A spacecraft offers a new means of charting shoals and shorelines."

"For centuries man has looked on the Nile Valley as one of the cradles of civilization," Astronaut FRANK BORMAN noted. "Generations have explored, excavated, and interpreted the significance of the Nile and its delta, but it was not until 1965 that the world received its first panoramic view of this sprawling spectacle on the northern coast of Africa. This picture revealed, for the first time as an entity, the 500 000-square-mile delta with its collar of wind-whipped rock and desert. This photograph became an important data point in man's quest to understand his environment."

"This photograph," says Robert O. Piland, Manned Spacecraft Center, NASA, "is one of an overlapping series extending from the Pacific coast of Mexico to central Texas. This frame shows the Colorado River entering the Gulf of California. Considerable geologic detail is apparent. The sinuous pattern in the gulf was originally identified as turbidity currents. The Naval Oceanographic Office has since demonstrated that this pattern is the bottom topography of the gulf, which is relatively shallow in this area. An orbital view of sediment distribution can be of great value to oceanographers."

"In the southwestern portion of the Rub al Khali, or Empty Quarter on the Arabian Peninsula," says Richard W. Underwood of the Manned Spacecraft Center, NASA, "we see the long, parallel 'seif' dune ridges which run unbroken for up to 400 miles and in some instances are 500 feet in height. The hollows between the ridges expose the bedrock and gravel flats. The background shows part of Arabia's Hadramawt Plateau. This portion of Arabia is poorly explored. Photographs which show such large portions of the Earth's surface provide information previously denied the geoscientist."

"An area approximately 70 miles on a side within the picture," says A. H. CHIDESTER of the Astrogeology Branch, U.S. Geological Survey, "is in the Sonoran Desert in Mexico. It encompasses a surprising range of climatic conditions, from the barren desolation of the Desierto del Altar to the lush vegetation of the 'arboreal desert.' Geologically the area is famous for its fine display of volcanic features. Complex volcanic mountains, cinder cones, wide craters, and flows are clearly visible in the photograph. Two of the maars have served as sites for training astronauts in field geology."

"The 22-mile-wide Richat structure in Mauritania is one of the most spectacular geologic features in Africa," according to PAUL D. LOWMAN, JR., of the Goddard Space Flight Center, NASA. "Resembling a gigantic bull's-eye, the Richat is a series of concentric ridges of resistant sedimentary rock. Structurally it is a domal uplift; its innermost rocks are Lower Cambrian limestones, surrounded by progressively younger rocks. The origin of this immense dome is not clearly understood. This synoptic view of the entire Richat structure and the surrounding area may throw new light on its origin."

"The Florida Keys and the tip of the mainland were photographed as part of an experiment in synoptic terrain photography," says CHRISTOPHER C. KRAFT, Director of Flight Operations, Manned Spacecraft Center, NASA. "This photograph represented a new era in the study of the Earth and its resources, just as Gemini IV represented the beginning of a new era in long-duration manned space flight." The Everglades are in the upper left. An angular line in the upper center is a road leading to Key Largo. Shoal areas from there to Boca Chica Key are visible. The whitish area is Sun glitter on the water.

Men Stepped Out in 1965

Cosmonaut Alexei Leonov (above) stepped outside of the Soviet Voskhod II for man's first "walk" in space on March 18, 1965. His command pilot, PAVEL BELYAYEV, photographed him and recalled later:

"Having satisfied myself that all of Alexei Leonov's life-support systems were functioning normally and his pulse and respiration were normal, I, at the prescribed time, gave him the order to emerge into outer space. During the period when Leonov was in outer space, I followed all his actions using a TV installation for this purpose. The TV camera operated very well and clearly. I saw in detail all the actions conducted by Leonov in outer space. I was in constant communication with him by telephone, and the instruments in the cabin helped me to check the functions of his individual life-support system. All the movements along the spaceship's outer shell as well as pushes could be heard distinctly inside the ship. Thus, besides all the envisaged systems for control of the cosmonaut's movements in outer space, there appeared an additional sound system of control. The cosmonaut's entry into the spacecraft was carried out in precise accordance with the program."

Astronaut Edward H. White (above and on the facing page) was the first American to leave the cabin of an orbiting spacecraft. He ventured out of Gemini IV on June 3, 1965, and his command pilot, JAMES A. MCDIVITT, wrote of the picture above:

"The hand-held maneuvering unit in his left hand was used for maneuvering near the spacecraft, while the camera mounted on top of it was used for EVA (extravehicular activity) photography. The chest-pack strapped to the harness received oxygen from the spacecraft through the gold umbilical and supplied oxygen to the suit for both pressurization and breathing.

"The flight helped to demonstrate the feasibility of EVA and the usefulness of a maneuvering unit when not in contact with a spacecraft."

Extravehicular activity later proved somewhat more difficult than was anticipated, and Astronaut Edwin E. Aldrin, Jr., who is shown engaged in it on pp. 183 and 184, used body tethers and handholds. With these, he successfully obtained experience in the performance of basic tasks with electrical and fluid connectors, hook-and-ring combinations, plastic strips, and fixed and removable bolts.

Astronaut EDWARD H. WHITE's recollections were as follows: "This was a picture taken by my teammate, James A. McDivitt, on the third revolution of Gemini IV. I had a specially designed spacesuit which had 21 layers of thermal and micrometeoroid protection. My face was protected by a double gold-plated visor which provided protection from the unfiltered rays of the Sun. In my hand I held a small self-maneuvering unit which gave me control of my movements in space. On my chest was an oxygen chestpack that regulated the flow of oxygen to my suit and provided an 8-minute supply of emergency oxygen. I was secured to the spacecraft by a 25-foot umbilical line and a 23-foot tether line, which were secured together and wrapped with a golden tape for thermal insulation. On the top of the hand-held self-maneuvering unit was mounted a 35-mm camera to record the event from outside the spacecraft."

Astronaut White died 2 years later, with Astronauts Virgil Grissom and Roger Chaffee, when fire swept the interior of an Apollo spacecraft at Cape Kennedy.

A Variety of Shores

Flight durations grew longer in 1965. In August, Astronauts L. Gordon Cooper and Charles Conrad, Jr., circled the world 120 times, in 190 hours and 55 minutes, and returned with 250 photographs in color. They used a NASA-modified hand-held Hasselblad, and both Ektachrome MS (S.O.–217) and Anscochrome D–50 film.

The nine examples of their work which follow show areas where Earth's lands and waters meet. All of the Gemini flights were within a band extending from 30° N to 30° S. Hence none of these portraits of the Earth's surface includes the difficult-to-explore regions near the poles. Even so, they show great variations in the contour of the surface and demonstrate the advantages of space photography in both thinly and densely populated areas.

Like many other Gemini photographs, the one on the facing page encompasses a broad span of scientific disciplines. For the meteorologist, there are cirrocumulus cloud formations indicative of the wind activity off the coast of California. The coastline is well defined, and the absence of sediment offshore is characteristic of a dry summer that resulted in little runoff. The haze layer covering Los Angeles is typical of the season in that wind patterns aided by the surrounding mountain ranges tend to amplify that area's smog situation. Farther inland, a broad sweep of the mountain ranges and the desert beyond are visible in fair detail for the geologist.

The other pictures in this section include new and spectacular views of the first part of the New World to be described by Europeans, the Bahama Islands and the peninsula of Florida. These are followed by one of islands in the Mediterranean that have

figured prominently in men's affairs for centuries, three pictures of Africa, and two more taken over the Pacific Ocean.

"As pressures on natural resources increase, because of growing populations and rising standards of living," Robert N. Colwell of the University of California at Berkeley told *Scientific American* readers in January 1968, "it becomes steadily more important to manage the available resources effectively. The task requires that accurate inventories of resources be periodically taken. Until as recently as a generation ago, such inventories were made almost entirely on the ground. Geologists traveled widely in exploring for minerals; foresters and agronomists examined trees and crops at close hand in order to assess their condition; surveyors walked the countryside in the course of preparing the necessary maps. The advent of aerial photography represented a big step forward."

Photography from orbiting platforms represented a further advance. So, too, did improvements in cameras and the development of additional remote sensing apparatus. Many bands of the electromagnetic spectrum outside of the visible-light portion now can be used to obtain information to supplement that provided by conventional photography.

Gemini V photography greatly encouraged the Earth Resources Survey program in which NASA is participating with such user agencies as the Department of Agriculture, the Department of Commerce, and the Department of the Interior. The program's overall objectives are to determine those Earth resource data that best can be acquired from space, and to develop further man's technological capabilities.

152

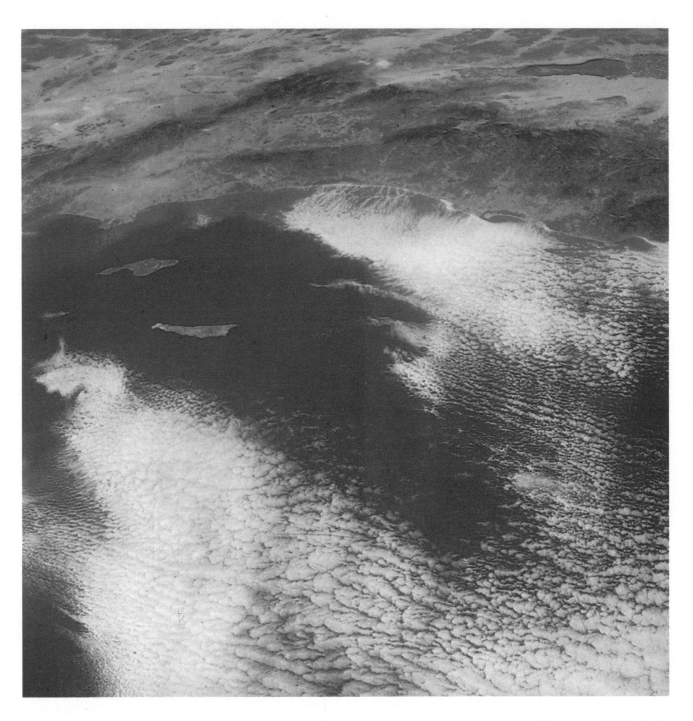

"It was 3 minutes past 11 o'clock in Los Angeles, on the far left, when this photograph was taken," says Samuel H. Hubbard, then Special Assistant, Gemini Programs, NASA. "San Clemente Island and Santa Catalina Island resemble steppingstones leading to Long Beach. The coast is visible from about Malibu to Bahia del Descanso, Mexico, the indentation on the extreme right. The inlet right of center is San Diego Bay, and the mountain ranges around Los Angeles are in the background. Behind them lies the Mojave Desert. The Imperial Valley and the Salton Sea are in the upper right."

"Clearly visible here," KURT H. DEBUS, Director of the John F. Kennedy Space Center, points out, "is the peninsula which has been the setting for all NASA manned flights and many of the unmanned scientific space exploration missions. Launch complexes, road networks, and the causeways connecting Cape Kennedy and the adjacent John F. Kennedy Space Center with mainland Florida can be readily identified. The Gemini photography demonstrated the capabilities of photographing Earth landmarks from an orbiting spacecraft. Succeeding manned missions employed these photographs for reference."

154

"This striking photograph has done much to galvanize the thinking of scientists to the possibilities of doing oceanography from space," says Arthur Alexiou of the National Science Foundation. "The lower left shows the Tongue of the Ocean, a deep-ocean reentrant incised into the Great Bahama Bank. To the east are Exuma Sound (a similar channel) and the Cat, Great Exuma, Long, and Eleuthera Islands. The sharp color demarcation hints the channel walls must be very steep. The unexpectedly clear definition of bottom topography suggests use of space photography for mapping it."

The clockwise stratocumulus cloud spiral at the left was northwest of Ifni, a Spanish possession surrounded by Morocco. Inland are 7700-foot peaks of the Anti-Atlas Mountains and the Jabel Quarkziz. "The spiral," says NORMAN G. FOSTER, Manned Spacecraft Center, NASA, "represents a lee eddy in-duced by the airflow around Cape Rhir to the north. Tals Mountain peaks also influenced the formation. Tiros X photographed this area about an hour later. A detailed comparison of the two views was published by the Department of Commerce in the May 1966 *Monthly Weather Review*."

156

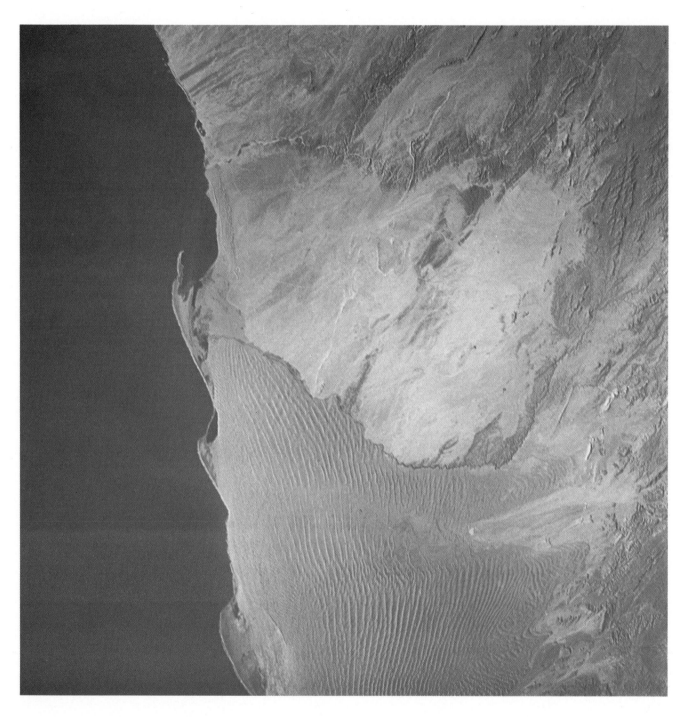

"This picture of southwest Africa covers a complex geological and geographical area that has not been fully exploited but that is of proven economic importance," says RICHARD J. ALLENBY, Deputy Director, Manned Flight Experiments Office, NASA. "The northern flowing Bengula current has shaped the encroaching land sands into a series of great hooks. Walvis Bay, the chief port of southwest Africa, is inside the northernmost sandhook, protected from the Atlantic by 5-mile-long Pelican Point. A strip of the coast extending south out of the picture is a restricted diamond mining area."

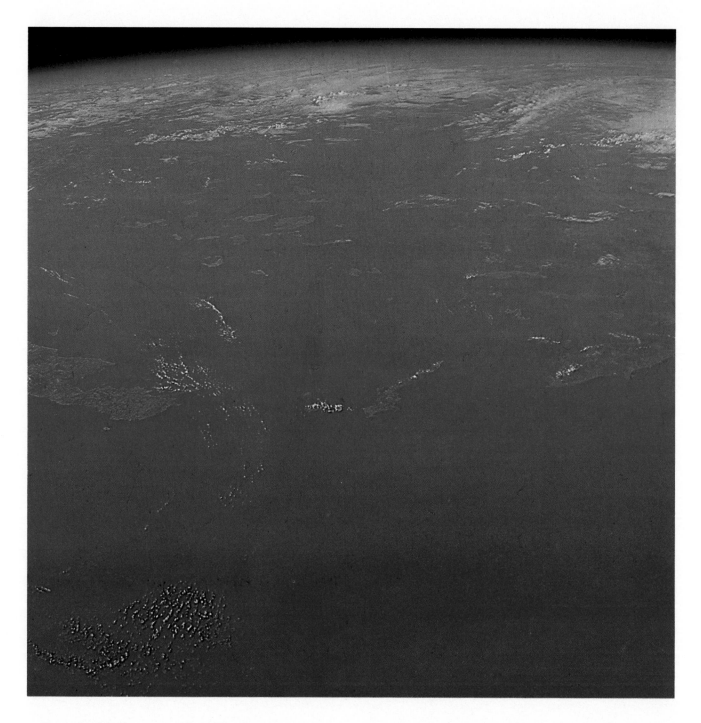

Over the Mediterranean Sea this photo was taken. "The islands of Crete, Carpathos, and Rhodes are in the center foreground from left to right," says JOHN R. BRINKMAN, Chief, Photographic Technology Laboratory, Manned Spacecraft Center, NASA. "The view is toward the north-northwest looking into the Aegean Sea. The Cyclades Islands are in the upper-left center; the northern Grecian mainland is under the cloud cover in the upper center. The coast of Turkey can be seen on the extreme right. One could, if desired, pinpoint renowned and ancient battlefields in this picture."

This view of the Nile Delta covers nearly 500 square miles. "Gross land-use patterns are apparent," notes Arch C. Gerlach of the U.S. Geological Survey. "Dark, irrigated croplands contrast sharply with the desert. Light spots show the extent and shape of nearly 400 populated places, of which the more conspicuous are Cairo near the delta apex (in lower center), Banhā and Zifta-Mit Gamr along the Nile to the north, and Tantā in the upper left corner. The apparently trifold distribution of urban centers has theoretical importance as a support for Christaller's central place theory applied to deltas."

"Honshu, the largest of the four main islands of Japan, is shown here with the Pacific Ocean in the foreground and the Sea of Japan in the background," reports one of the principal investigators for the Gemini synoptic terrain and weather photography experiments, PAUL D. LOWMAN, JR., of the Goddard Space Flight Center, NASA. "Osaka, a city of more than 8 million, is situated at the far left on Osaka Wan (bay). The port of Nagoya is at the upper right edge of Ise Wan in the foreground." Honshu is a mountainous island with a coastline that has both steep cliffs and lowlands with fine harbors.

160

"Here the island of Hawaii with its volcanic peaks of Mauna Loa and Mauna Kea is veiled by Pacific clouds," STANLEY SOULES of the National Environmental Satellite Center has pointed out. "Partly hidden by the clouds on the right are the islands of Maui, Kahoolawe, Lanai, Mokokai, and Oahu."

Populated first by men navigating by the stars in double-hulled canoes, the State's many islands and islets extend across the Pacific for 1600 miles and have figured increasingly in world affairs. Mark Twain called this "the loveliest fleet of islands that lies anchored in any ocean."

There are 34 atolls and hundreds of coral reefs in the Marshall Islands. "This picture," says Willis B. Foster, Director, Manned Flight Experiments Office, NASA, "includes all of Rongelap (in center), Alinginae (at top), and a tip of Rongerik (at bottom). Rongelap is about 60 miles long. All three are typical North Pacific atolls, consisting of an annular reef, probably lying on top of a truncated volcano. Islands along the reef can be built, moved, or destroyed by a single storm or by subsidence of the volcano. Photos such as this will add immeasurably to study of their formation, life, and death."

Guadalupe Island, Mexico, is in the foreground and Baja California in the background of this photo of a double vortex in stratocumulus clouds. "Such vortices," says KENNETH M. NAGLER, Chief, Space Operations Support, Environmental Science Services Administration, "are produced when the wind flows past the mountain island (a peak of 4257 feet) in much the same way that eddies are induced by an obstruction in a flow of water. Although these vortices are not related to severe weather, their formation and movement is under investigation—largely to determine what they reveal about the atmosphere."

Rendezvous at 17000 mph

In December 1965, two Gemini spacecraft were aloft at the same time. In Gemini VI, Walter M. Schirra, Jr., and Thomas P. Stafford went around the world 15 times, and in Gemini VII, Frank Borman and James A. Lovell, Jr., completed 206 revolutions —taking more pictures, frequently with the same kind of Hasselblad used on earlier flights, from altitudes that ranged up to 200 miles.

Gemini VI was piloted into the same orbit as Gemini VII and the two craft remained together for more than 5 hours, at times only a foot apart.

"My first indication," Astronaut WALTER M. SCHIRRA, JR., wrote of the picture below, "was that there was an exceptionally bright star out there, probably Sirius. A good look at the constellations in the Sirius and Orion area indicated no bright star at that location. It had to be spacecraft VII. It disappeared like a light going out when it went behind the terminator at 43 miles' range. When we both came into the sunlight again, we were at rendezvous. On the next revolution, Tom [Astronaut T. P. Stafford] took this photo."

GEMINI VI

The picture at the right was taken from Gemini VI of Gemini VII while they were about 13 feet apart on December 15, 1965. Astronaut THOMAS P. STAFFORD remembers: "The ability to take this photo was very gratifying after the numerous delays we encountered in performing the world's first rendezvous mission. The Gemini VI spacecraft was launched on the third attempt to perform the rendezvous and performed it flawlessly. After the hard work and effort of over a year, we finally achieved the key step that will lead us to the lunar-landing mission. The sight was utterly fantastic to fly in close formation as we have done for many years in fighter aircraft, but at 17 000 miles an hour. The photo illustrates that man can control a space vehicle with preciseness in close vicinity to another vehicle."

"One of the tasks of Gemini VI was to inspect the condition of Gemini VII after 12 days of space flight," Astronaut JAMES A. LOVELL, JR., wrote as a caption for the lower picture here. "The photograph shows the operation being conducted along the right side of Gemini VII. Clearly visible are the atmospheric heating effects in the paint of the words 'United States' and in the U.S. flag. In the center of the spacecraft can be seen the buildup of frozen urine particles on the triangular red-rubber structure accumulated after 12 days of dumping. On the white adapter area just behind the shadow of the infrared sensor door are scattered particles of frozen fluid vented from that device. A close examination of the window will reveal me peering anxiously out of the corner as Gemini VI disappears behind us. The rendezvous of Gemini VI with Gemini VII was a milestone in space flight. It was the first opportunity to closely examine a space vehicle in orbit and proved our concept of rendezvous."

"The initial impression is that this photograph indicates a tropical-storm-cloud system," KENNETH S. KLEINKNECHT, Deputy Manager, Gemini Program, Manned Spacecraft Center, NASA, says of this picture in which Baja California and the gulf between it and Mexico are discernible. "Closer examination reveals a cloud mass associated with a low-pressure center. The large, circular cloud pattern in the center is thick altocumulus. Gemini photographs have given men a clearer understanding of weather phenomena, as well as information to aid in interpretation of weather-satellite photographs."

This "majestic roof, fretted with golden fire" lay below Gemini VII as it crossed South America. "The low angle of the setting Sun produced this unusual glow of clouds along the Andes Mountains," says LEROY E. DAY, Apollo Program Test Director. "This picture looks south from northern Bolivia across the Andes Mountains, and illustrates the extensive cloud cover seen in that area. The Gemini VII mission produced a wealth of medical, scientific, and operational data which clearly established man's ability to work usefully during a 2-week space flight and return to Earth in good health."

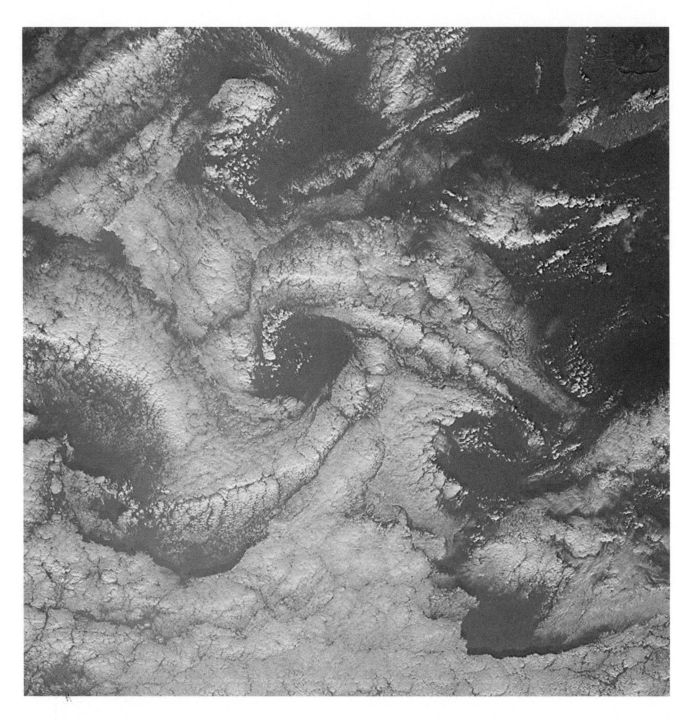

"When northeast winds blow past the mountainous Canary Islands," says STANLEY SOULES of the National Environmental Satellite Center, "they frequently are swirled into eddies that show up in the stratocumulus clouds. The 12 198-foot cratered mountain on Tenerife is in the upper right, and Gomera, partly hidden by clouds, is farther left. The eye of the center eddy is about 13 miles across. Two other eddies rotating clockwise are near it. The cloud tops are about 3500 feet high. A temperature inversion exists at this altitude and a chain of eddies analogous to a Kármán vortex street has formed downwind."

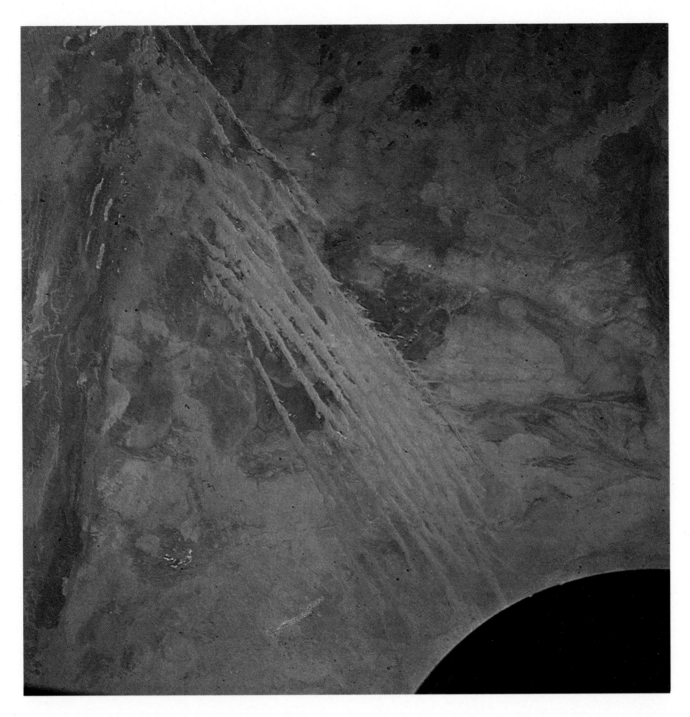

Natives call this harsh, lifeless plain in Africa's western Sahara the Tanezrouft, or "Land of Terror." PAUL E. PURSER, Special Assistant to the Director, Manned Spacecraft Center, NASA, says: "The long stripes from northeast to southwest are 'seif'-type sand dunes. The sand, obviously blown in from a distant area, is slowly migrating across the bedrock. Photographs covering such large areas provide a great deal of data on such regional migrations of sand dunes and their geomorphologic aspects. The continuity of features and their gradations are often masked in mosaics of low-altitude photographs."

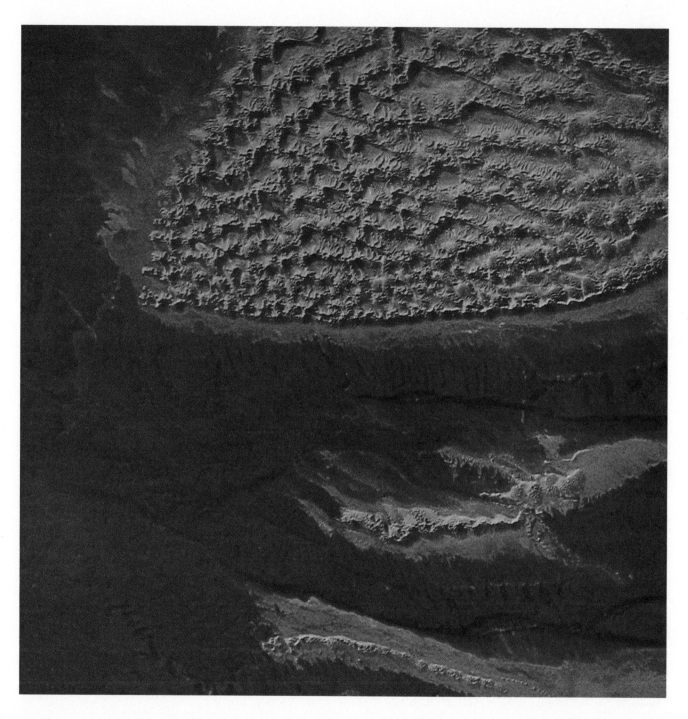

"This desert photograph of North Africa," says Astronaut JAMES A. LOVELL JR., "is typical of the beautiful but desolate views this part of the world offers to the space traveler. The area was almost always free of clouds and a delight to photograph because it presented a vast array of brilliant colors without haze obstructions. The photograph dramatically shows the results of wind and water erosion. For hundreds of miles one could see row after row of sand dunes surrounded by deeply scarred mountains. It was truly the most interesting area of the world to view from space." This field was in Algeria.

170

"This unprecedented synoptic view of Lake Chad in Africa," says WILLIAM T. PECORA, Director of the U.S. Geological Survey, "illustrates a major contribution to broad-scale studies of hydrology in less-developed countries. The remarkable transitional nature of the eastern shore is easily seen. The land consists of sand dunes of various heights and configurations, and there is a gradual change from dry land to land with ponds, to land with bays, to bays with peninsulas and islands, to islands, to open lake. The westward limit of the submerged dunes shows the extent of the lake during past dry periods."

Targets Attained

Another key event came March 16, 1966: The first dual launch and docking of a manned U.S. spacecraft with an unmanned target vehicle. The astronauts were Neil A. Armstrong and David M. Scott. The picture below was taken when the target was about 45 feet from the nose of their spacecraft (at lower left). "This Agena target vehicle," ARMSTRONG notes, "was the first unmanned satellite successfully photographed from space. The picture was taken over South America by David Scott, just prior to the first successful docking of two spacecraft on the flight of Gemini VIII. It clearly indicates the detail in which one satellite can be observed from another. This photograph is a particularly good replica of the actual view seen with the eye, with the exception of the brilliance of the white and metallic parts of the Agena, never yet captured on film." A short circuit in the orbital attitude maneuvering system that bled fuel through one thruster necessitated curtailment of the Gemini VIII flight and the astronauts descended to a preplanned emergency landing area in the Pacific.

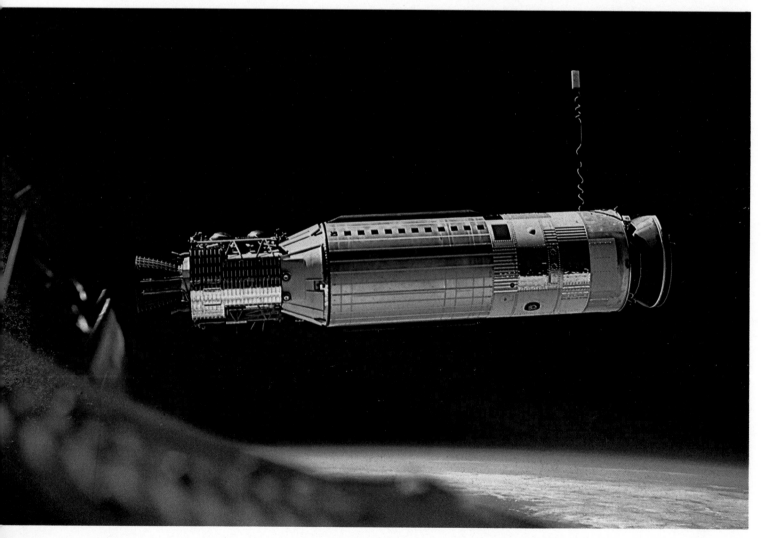

A volume of memorable Gemini photographs would be incomplete without this one taken on June 3, 1966. WILLIAM C. SCHNEIDER, who is now Mission Director of the Apollo Program, explains this startling sight in this way: "Four hours and 8 minutes after liftoff, Astronauts Tom Stafford and Eugene Cernan had their first view of their rendezvous target, 'the angry alligator.' The shroud covering the target, the augmented docking adapter, had not separated and was held by a restraining band. Docking of the two vehicles was impossible. After a careful inspection and report on the conditions, the contingency flight plan was initiated and the mission continued. Two additional rendezvous, two extravehicular exercises, and five scientific and technical experiments were performed prior to a near-perfect reentry and landing."

JAMES ELMS, former Deputy Associate Administrator for Manned Space Flight, who now directs NASA's Electronics Research Center, continues the account of that 1966 flight of Gemini IX in this legend for the photo below on this page. The picture was taken by Cernan from the end of the umbilical. "After successfully completing their assigned rendezvous exercises, Astronauts Tom Stafford and Gene Cernan elected to conduct the first of two scheduled extravehicular periods on the third day. Cernan first performed several tasks involved with the evaluation of tether dynamics—notice the 'rearview' mirror attached to the docking bar and the movie camera attached beyond the hatch. Later Gene moved to the rear of the spacecraft where he attempted to don the astronaut-maneuvering unit, a back-mounted 'personal rocket.' Unfortunately, this task was harder to accomplish than anticipated and this extravehicular exercise was terminated."

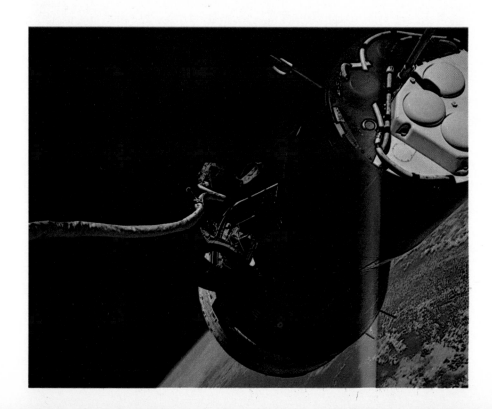

"Here we are checking the various systems on our Gemini X Agena target vehicle prior to lighting its large engine for the first time," Astronaut JOHN W. YOUNG reports regarding this picture. "The Agena is also controlling our attitude, a convenient way of keeping us level, but one which doesn't allow a very good look at the world below. That little piece of blue and white in the lower corner of the window was all the scenery we saw for a day and a half."

Above

and Beyond

In the summer of 1966, Gemini IX and X left Cape Kennedy to continue rendezvous, extravehicular activity, and other experiments. Astronauts Thomas P. Stafford and Eugene A. Cernan completed 44 revolutions aboard Gemini IX, and John W. Young and Michael Collins completed 43 on Gemini X. The latter docked their spacecraft with an Agena target vehicle and then used its propulsion system to shift them into an orbit that reached an altitude of 475 miles.

Their cameras included not only the NASA-modified Hasselblad Model 500–C used successfully on previous flights, but also super-wide-angle Hasselblads with Zeiss Biogon 38-mm lenses and 70-mm space cameras with Xenotar 80-mm lenses.

The pictures on the next few pages include examples of the photos they obtained. The first one is of the Mediterranean's western portals. The second one is a photo taken later of the Gattara Basin in the United Arab Republic which has been considered as

a possible electric power site since 1927. Here some 5130 square miles are about 164 feet below sea level. Replacement of water lost by evaporation from a lake formed in this depression could be used to generate power. The proximity of the Mediterranean offers an unlimited water supply. Photographs such as this from space can be extremely helpful in planning and developing hydroelectric power systems.

Views of thinly populated parts of Africa and North America follow, and the final photo in this group was taken on what Astronaut Cernan insists was "the most fascinating and beautiful trip a man ever made across South America." His course was one that would be arduous to follow on wheels. "Without blinking an eye," he says, "I could see the high Andes, Pacific Ocean, the great Altiplano with a jewellike Titicaca, the rain forests of the Amazon basin, and the Chaco plains on down our orbital path."

174

"Europe and Africa enjoying good weather—but not for long if that storm off Gibraltar is any indicator," MICHAEL COLLINS thought when he took this picture. "This is about as far north as we went on Gemini X," he noted later, "since we launched nearly due east from Cape Kennedy." Portugal and Spain are in the upper left, and Morocco is at the right. Although the range keeps one from seeing many details, the essential geologic unity of southern Spain and Africa is suggested in this photo by the obvious continuity of the Sierra Nevada and related mountains in Spain with the Riff Atlas in Morocco.

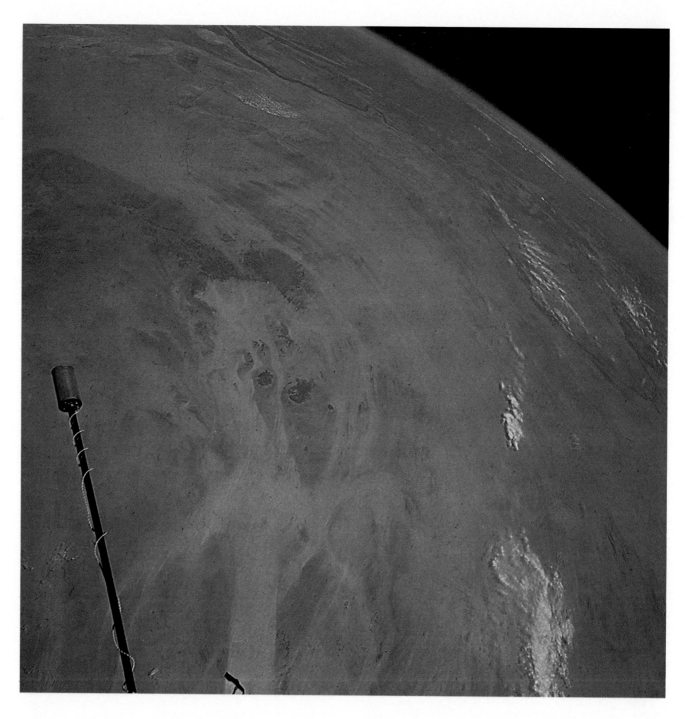

Over north-central Africa, looking northeast, Libya, the United Arab Republic, and the Sudan could be seen. On the horizon are the Arabian Peninsula, the Red Sea, Nile Delta, and Gulf of Suez. "The prominent dark circular objects toward the center," says JACOB I. TROMBKA of the Goddard Space Flight Center, NASA, "are hills believed to be granitic extrusions. Beyond and above them is the Gilf Kebir Plateau. Just above this plateau is the Great Sand Sea. Its dunes rise 150 to 300 feet. The change in colors can be attributed to the change from sand to rock and rock debris."

176

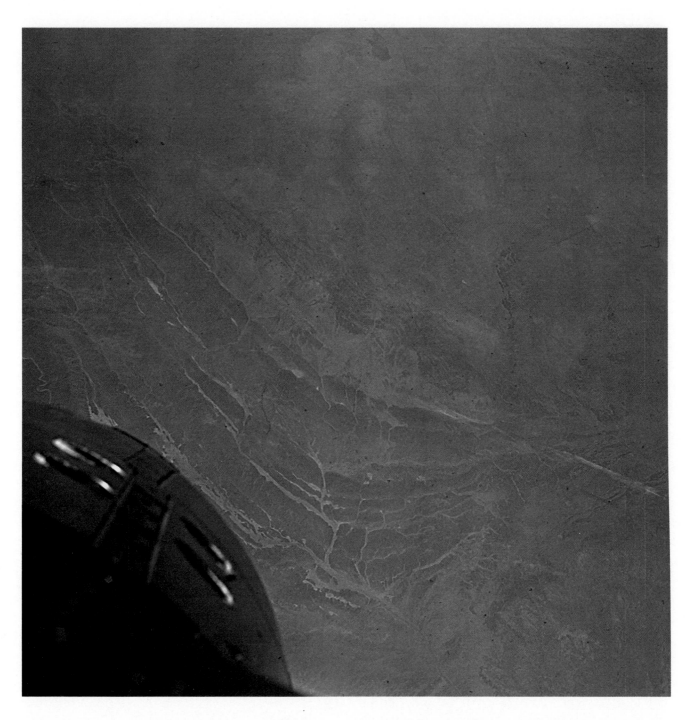

"This photograph," says Wesley L. Hjornevik, Director of Administration, Manned Spacecraft Center, NASA, "covers portions of Mauritania, Algeria, and Spanish Sahara to the extent of about 50 000 square miles, or about the size of Illinois. At the left in the Yetti, or 'White Plains,' numerous intrusive volcanic dikes can be seen. In the foreground are the long folded ridges of the Mechems, and the close relationship of the wadi (dry stream bed) system to the geologic rock structure is evident. This photograph provides a great deal of information about a remote and relatively unexplored region."

"Gene Cernan recorded this clear view of Baja California and western Mexico," says Astronaut ALAN B. SHEPARD, JR. "The view along the spacecraft nose is generally toward the southeast. The local time is midmorning with the Sun shining from the left. Because the spacecraft hatch was open, just visible at the right, unusual clarity exists in this photo. Although the offshore wind pattern is not noticeable without clouds, there is a fine example of vortex flow off Cabo de San Lucas, the southern tip of Baja California. Excellent study of drainage patterns is apparent in the foreground on the peninsula."

"This sweeping view southwestward," says JAMES F. SEITZ of the U.S. Geological Survey, "shows the Andes Mountains and Altiplano Plateau of Bolivia, the coasts of Peru and Chile, and the Pacific Ocean. Lake Titicaca in the foreground is the highest lake in the world navigated by steamships; 300 miles south, salt flats cover more than 5000 square miles. Bordering the Altiplano on the east (lower foreground) are 21 000-foot, snow-capped peaks. The great canyons cut into the Altiplano show where gaps between granitic bodies have permitted erosion. Thus the distribution of granite is revealed."

Zodiacal Light

The glow we see briefly in the east before dawn, and in the west after twilight, is the zodiacal light. Scientists have long sought information about our solar system in this light. The picture here was taken for them above the atmosphere. The bright spot at the top is Venus. The zodiacal light is shown between Venus and the thin airglow band above the moonlit Earth.

EDWARD P. NEY of the University of Minnesota points out this picture's uniqueness: "This photo was taken by Astronauts Stafford and Cernan on June 4, 1966, 4 minutes before satellite sunrise with a field of view of 130° by 50°. At the bottom of the picture is the Earth, illuminated by a nearly full Moon. The thin horizontal line 2° above the Earth's limb is due to the airglow layer. This is a self-luminous region of the Earth's atmosphere extending from 80 to 100 kilometers' altitude. The cone of illumination above the airglow layer which seems to point at the bright planet Venus is the phenomenon of zodiacal light. It is symmetrical about the ecliptic, which is inclined to about 45° with respect to the horizon in this picture.

"The zodiacal light is produced by sunlight scattered from dust grains in orbit about the Sun. The unique aspects of the space photograph are that the airglow layer is seen as a line because the spacecraft is above the airglow layer, and the zodiacal light is not distorted by the Earth's atmosphere as it is when viewed from the ground."

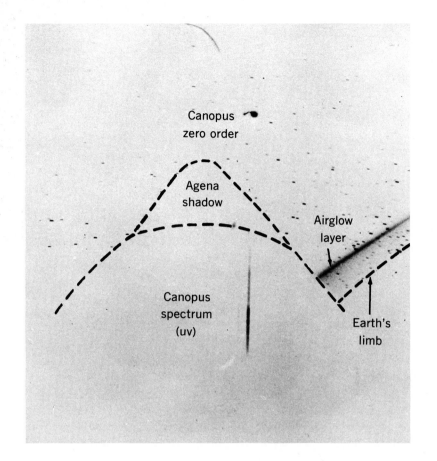

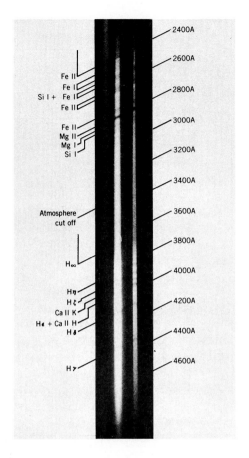

And a Star's Spectrum

During two hours of extravehicular activity on Gemini XI, Astronauts Conrad and Gordon photographed ultraviolet stellar spectra in six regions of the sky. The camera was attached to the spacecraft frame; Gordon stood up in the open hatch to operate it while Conrad remained in his seat to control attitude and time the exposures.

Of the photograph of Canopus above, astronomer KARL G. HENIZE of Northwestern University, who is now himself an astronaut, writes:

"The numerous short horizontal lines are zero-order images of background stars which are elongated due to a slight yaw motion of the spacecraft. The vertical streak directly below this bright zero-order image of Canopus is its spectrum. The lower portion of the spectrum shows lines due mainly to the familiar Balmer series of hydrogen. At the upper end appear several newly observed ultraviolet absorption lines, the strongest of which is the ionized magnesium doublet at 2800 Å. Other features are due to silicon and to neutral and ionized iron.

"This photo was obtained with the 73-mm focal-length, $f/3.3$ lens of the Maurer camera, to which a 600 line/mm diffraction grating was attached. The exposure time was 20 seconds.

"This spectrum is the first to show absorption lines in the 2200- to 3000-Å wavelength region of a star other than the Sun."

Tethered Targets

From both Gemini XI and Gemini XII, an Agena target vehicle was photographed while tethered to the spacecraft. Astronaut RICHARD F. GORDON, JR., says of the picture of himself at work at the right: "This photograph shows me astride the nose of the Gemini spacecraft during the 'umbilical EVA' of Gemini XI. This pose stimulated Pete Conrad to respond with the words, 'Ride 'em, cowboy!' The task involved was to attach a 100-foot Dacron tether to the docking bar on the Gemini spacecraft. This was the first time that two objects had been tethered together in space."

In the lower picture on this page, says JOHN A. EDWARDS, Operations Director, Apollo Applications Program, NASA: "The Agena target vehicle is shown tethered to Gemini XI. The 100-foot Dacron tether, with one end attached to the spacecraft docking bar and the other to the target docking adapter, was still in the process of extension at the time this photograph was taken, and the two vehicles are positioned such that the astronauts are looking directly into the docking cone. Below the vehicles is the Gulf of California and Baja California at La Paz.

The objective of this operation was to evaluate the feasibility of station keeping two vehicles in space by gravity-gradient stabilization or by slow rotation of the tethered vehicles, thereby setting up an artificial gravity field. Because of initial difficulties in establishing a gravity-gradient-stabilized system, the two vehicles were 'spun up' by Command Pilot Pete Conrad shortly after this picture was taken, and 2 hours and 46 minutes of station keeping was maintained at angular rates of 38 to 55 degrees per minute."

"Working hard at the Agena work station, 'Buzz' Aldrin did not know when I took this picture of him through the left-hand window of Gemini XII," reports Astronaut JAMES A. LOVELL. "It occurred during his second trip outside and 'Buzz' was already establishing new milestones in man's ability to complete useful tasks in the void of space. The successful completion of the 5½ hours of extravehicular activity during the Gemini XII mission helped to pave the way for future space missions."

Astronaut EDWIN E. ALDRIN, JR., writes: "This photograph was taken during Gemini XII on November 12, 1966, during the first standup EVA. The camera used was a 16-mm Maurer movie camera which I had installed just after sunrise on the retroadapter aft of the open right hatch. The camera was actuated by the command pilot, Capt. James Lovell, using a remote cable assembly. The experiment package shown is an SO–12 micrometeorite collector, which also was mounted aft of the right hatch and had been opened and closed remotely from the ground for exposure during the first sleep period. Prior to removal, the blue tether with a brass hook was attached to the experiment handle. The photograph shows me in the process of temporarily stowing the experiment in the empty food pouch shown on the open hatch under my left elbow."

CHARLES MATHEWS, then Gemini Program Manager at Manned Spacecraft Center, NASA, points out: "This photograph of Gemini XII pilot Edwin E. Aldrin, Jr., was taken on November 12, 1966, with a 70-mm Hasselblad mounted on the spacecraft while Pilot Aldrin was standing in the open hatch. In the lower left is a rear view of the 70-mm general-purpose camera used for Experiment S–13 as an ultraviolet astronomical camera. The objective of S–13 was to record the ultraviolet radiation of stars, using an objective prism and an objective grating. An analysis of the surface temperatures of the stars, of the absorption effects taking place in their atmospheres, and of the absorption effects of the interstellar dust will be made from the photographic data obtained. In addition to the acquisition of basic astronomical data, techniques by which objective prism spectra may be best obtained were determined. The practical experience gained will be useful in planning similar astronomical observations with large telescopes on future missions."

FRANK A. BOGART, Deputy Associate Administrator for Manned Space Flight (Management), NASA, notes that: "This photograph was taken from Gemini XII above the Pacific. It shows the spacecraft tethered to the Agena target vehicle, with both maintaining a preferred orientation. This exercise demonstrated the feasibility of achieving stabilized orientation between two maneuverable spacecraft at slightly different altitudes, connected by a flexible tether. At the beginning of the exercise the two spacecraft were docked—the tether having been connected during Astronaut Buzz Aldrin's 'umbilical EVA.' The vehicles were pitched to a vertical stable attitude, undocked, and separated, extending the tether. The second attempt to 'capture' or establish the gravity-gradient-stabilization mode was judged successful in that the orientation was held for approximately 90 minutes with all control systems turned off. This successful demonstration promises a most useful technique for station-keeping spacecraft with minimum fuel expenditure."

GEMINI XII

The Astronauts' Record Climb

The Gemini flights were intended in part to determine the feasibility of sending men to the Moon, to perfect equipment and techniques to get them there and back to Earth safely. Photography of the Earth was but one of many experiments undertaken on those flights, yet the astronauts brought back more than 2400 photographs of the type reproduced here. Many of their pictures were both strikingly beautiful and enlightening to scientists, and many suggested further uses for manmade satellites of the Earth.

The equipment was improved as the program progressed and various films were used, most of which had emulsion coatings and bases especially formulated to meet NASA specifications. Flight films were sent through a processing machine singly, under close surveillance, and none was lost because of a laboratory malfunction.

In September 1966, Astronauts Charles Conrad, Jr., and Richard F. Gordon, Jr., photographed the Earth from an altitude of 741.5 miles, the farthest distance that men ever had gone from its surface. To attain this altitude, they docked Gemini XI with an Agena target vehicle and used its propulsion system to increase the apogee of their 27th orbit.

In December 1966, Astronauts James A. Lovell and Edwin E. Aldrin, Jr., concluded this historic series of flights by undertaking more extravehicular activity and successfully performing 14 scientific experiments—which included taking more pictures indicative of the value of regular surveys of our planet's resources from orbiting observatories.

Experience was gained on all of the Gemini flights with the variation of spacecraft orientation during reentry into the atmosphere. On Gemini VIII, unexpected roll-and-yaw motion after the spacecraft was docked to an Agena target caused the astronauts to undock and stop the motion with a special control system normally used only during reentry. Analysis revealed that a short circuit in the wiring of a control thruster caused the trouble. On the last two flights, a spacecraft computer controlled reentry after initiation by the astronauts.

"The Gemini XI mission was our 'moment of truth' concerning extravehicular activity," Dr. CHARLES A. BERRY, Director of Medical Research and Operations, Manned Spacecraft Center, NASA, recalls, "for Dick Gordon experienced severe fatigue while he was connecting a tether from the spacecraft to the Agena. This resulted in our careful planning to test man's capability in extravehicular activity on Gemini XII."

There were no manned space flights by Americans in 1967. The next 10 pictures were taken on the Gemini XI and XII flights. The launching pad fire, in which three astronauts perished, resulted in major changes in plans and equipment, and NASA's 10th anniversary was marked by further tests preparatory to a manned flight to the Moon.

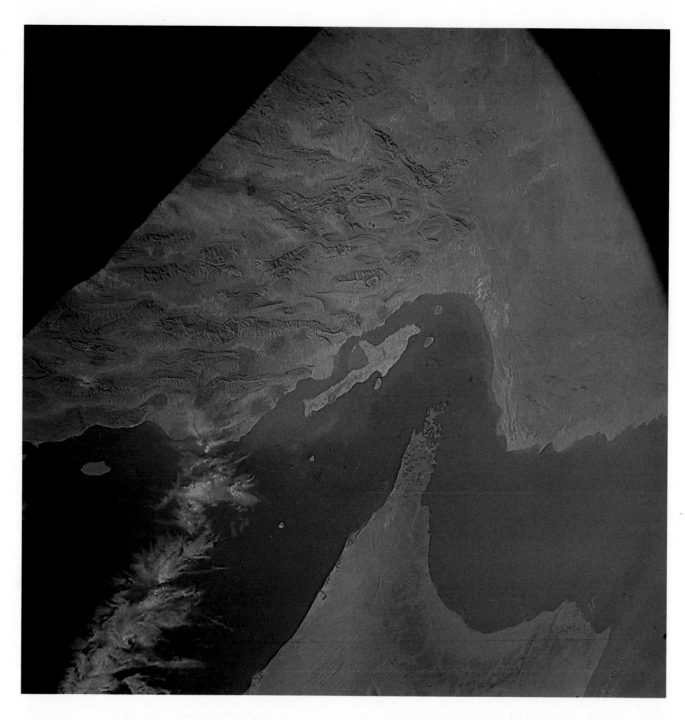

"Across the center of this study in blue, a 600-mile stretch of the gulf coast of Iran is visible," says DAVID M. JONES, former Deputy Associate Administrator for Manned Space Flight, NASA. "The picture is typical of the high quality of the Gemini photographs and includes usable geodetic and climatological information." The southern tip of the Zagros Mountain range is in the left center. The pinnacle of land in the lower center is part of the Arabian landmass. The Straits of Hormuz provide passage between the Gulf of Oman on the right and the Persian Gulf.

George E. Mueller, Associate Administrator for Manned Space Flight, NASA, notes: "The city of Houston appears in the left center of this photo of about 280 miles of the Texas-Louisiana gulf coast taken at an altitude of about 150 miles. Careful examination reveals Ellington Air Force Base and buildings of the Manned Spacecraft Center. The picture's significance lies in its depth of detail. The clarity of terrain features and of the flow patterns of each body of water emptying into the Gulf is outstanding. Pictures like this will in the future make possible a survey of Earth's resources from space."

A band of cirrus clouds crosses this picture of the Nile Valley and the Red Sea. "Such cloud bands," says KENNETH M. NAGLER, Chief, Space Operations Support Division, Environmental Science Services Administration, "normally occur on the equatorward side of the core of a strong westerly windflow known as the jetstream, in which winds of 100 knots to occasionally more than 200 knots are common. Often seen in less detail in weather-satellite views, these cirrus bands are particularly useful as indicators of the direction of the upper level wind. In the foreground small cumulus clouds are alined in rows."

"A photograph from the 190-mile altitude shows the beautiful color contrast between the Mediterranean and the sands of the Sahara," says EVERETT E. CHRISTENSEN, who directed manned space-flight missions for NASA at the time. "Close inspection of the circular sandy area, Idehan Marzūq, and the irregular sandy area to the left, Edeyin Ubari, will reveal dune formations. Dark areas in the upper right are mountains. While this is one of the world's least populous areas and the deserts are reported as having no rainfall, several million people reside on oases and in the mountain areas."

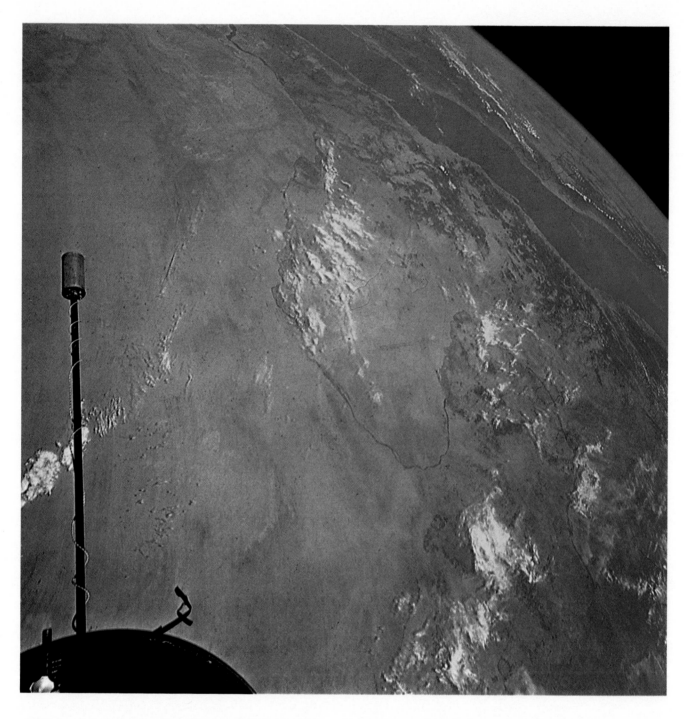

"The nose of the Gemini XI spacecraft with docking bar in lower left is in stark contrast to the orange of the Libyan Desert," DR. CHARLES A. BERRY of the Manned Spacecraft Center, NASA, points out. "The Nile traverses the desert like a carelessly dropped blue-green rope from Biba in Egypt (upper left) to Khartoum in the Sudan (lower right). The blue of the long gap across the edge of the world is the Red Sea separated from the blackness of space by Arabia. Astronauts Cooper and Conrad, our wives, and I flew over this area on a good-will tour following Gemini V."

190

"The success of the Gemini Program was due to the hard work of many elements, culminating in outstanding hardware and crew performance," says DONALD K. SLAYTON, Director of Flight Operations, Manned Spacecraft Center, NASA. "This photograph of the Near East is an example of how high-quality photographs can detect unusual features such as the oil fire in the upper right corner (a triangular dark spot), and thereby provide productive future applications for space photography." The Suez Canal is at the left, the Gulf of Suez in the foreground, and the Red Sea at the right.

JOCELYN R. GILL, Gemini Science Manager, NASA, observes: "The picture encompasses Sudan, Ethiopia, Somali, French Somaliland, Saudi Arabia, Yemen, and South Arabia. Since it is a high-altitude (340 miles) photograph, it contains less geologic detail than those taken at lower altitudes; however, for this very reason, the photo may be of special interest to the meteorologist for comparison with Tiros and Nimbus photographs." The Red Sea is in the foreground, the Gulf of Aden above it, and the Arabian Sea on the horizon. Saudi Arabia and the Empty Quarter show up clearly.

GEMINI XII

"This photograph from an altitude of 410 miles encompasses all of India, an area of 1 250 000 square miles," GEORGE M. LOW, then the Deputy Director, Manned Spacecraft Center, NASA, notes. "Bombay is on the west coast, directly left of the spacecraft's can-shaped antenna. New Delhi is just below the horizon near the upper left. Adam's Bridge between India and Ceylon, at the right, is clearly visible. A cloudless region surrounds the entire subcontinent. Differences in color, green near the west coast, and brown inland, delineate regions of heavy vegetation and semiarid areas."

GEMINI XI

"The altitude at the time of this photograph was 465 miles," according to MAXIME A. FAGET, Director, Engineering and Development, Manned Spacecraft Center, NASA. "The view is to the northeast and includes India and Ceylon at the horizon and the atolls of the Maldive Island chain near the photo-graph's center. To the west of India is the Arabian Sea, to the east is the Bay of Bengal, and to the south the Indian Ocean. The cloud structure had changed from that photographed on the previous revolution. Of particular significance are the high cumulus buildups over Ceylon."

GEMINI XI

Astronaut CHARLES CONRAD says of this picture: "The Sun was slowly settling into the west, causing an angular reflection in the glass window of the spacecraft. Dick Gordon and I were looking at western Australia. This picture was taken approximately when we reached the high point of our high-altitude orbit. This was the view from 739.4 miles. We were excited and although we had planned the high orbit for months, we never realized what a sight we would see. Fifty minutes earlier we had ignited our Agena rocket engine for the longest burn ever made to change an orbit."

The Next Big Step

One of the most portentous events of NASA's first 10 years occurred on November 9, 1967. Then, after nearly 6 years of intensive development, the Saturn V—developing 7 500 000 pounds of thrust—functioned perfectly on the Apollo 4 launch. The three-stage Apollo/Saturn vehicle is 363 feet high and is capable of placing nearly 140 tons in Earth orbit and of launching nearly 50 tons to the Moon.

The next three photographs increased confidence that men can examine the Moon's surface in person in the near future.

Stages of the three-stage Saturn V launch vehicle are discarded after burnout. The first stage, and the interstage separating the first and second stages, are jettisoned a few seconds apart in what is called a "dual plane" separation sequence. The photograph on the facing page here shows the interstage falling away from the second stage of the vehicle on the November 9, 1967, flight.

WERNHER VON BRAUN, Director of the Marshall Space Flight Center, NASA, explains what this picture shows and how it was obtained:

"The interstage is glowing from the intense heat of the invisible exhaust flames produced by the second stage's five J–2 engines, two of which are visible in the foreground, burning liquid hydrogen fuel and liquid oxygen. The first stage, which was separated about 20 seconds earlier, is visible far below as a small white shape just inside the right ring of the interstage.

"This picture was taken from a frame of 16-mm motion-picture film exposed during the dual plane separation. Two cameras were inside special water-tight capsules mounted on the second stage to record the action. The camera capsules were ejected a few seconds after this picture was taken at an altitude of about 40 miles. The capsules reentered the atmosphere and, slowed by paraballoons, splashed down in the Atlantic Ocean where they were retrieved quickly by recovery forces stationed in the area. Personnel were guided to the floating capsules by radio and light beacons and by dye released after splashdown. The cameras were flown back to land and rushed to the photographic laboratory at Marshall Space Flight Center for processing."

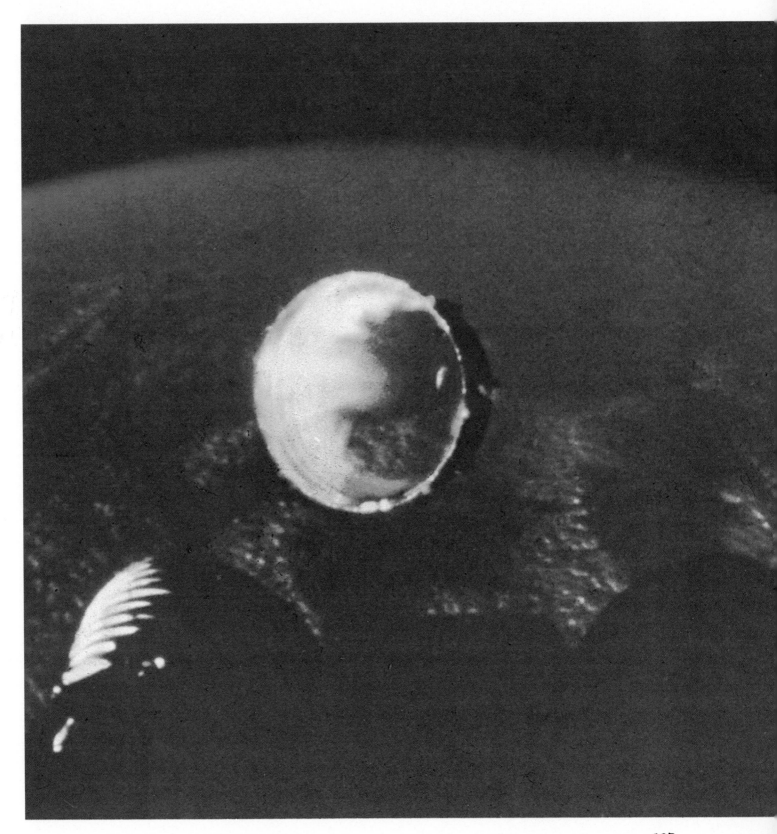

These pictures were taken through the command pilot's window on the unmanned Apollo flight November 9, 1967, while the spacecraft soared some 11 000 statute miles above the Earth.

"This is a sight astronauts will see on the way to the Moon," SAMUEL C. PHILLIPS, Director of the Apollo Program, NASA, commented.

The picture above on this page was taken when the spacecraft was over South Africa and so oriented that the camera viewed the area to the northwest.

"Viewing this picture from right to left," Phillips pointed out, "we see the Sahara Desert (western Africa), the east coast of South America on the left horizon, and a clearly distinguishable circular airmass flow around a South Atlantic depression. This airmass is commonly known as the 'Roaring Forties' which, in the days of sailing vessels, made the voyage between Cape Horn and the Cape of Good Hope quite difficult. Antarctica is not depicted on this particular photograph, but some ice floes are discernible.

"This photograph was one of 755 exposures made at 10.6-second intervals beginning 4 hours 28 minutes after liftoff. Using Eastman Kodak Ektachrome MS, type S.O.–368 film, it was taken by a Maurer Model 220G 70-mm sequence camera 5 hours and 6 minutes after the liftoff."

The photograph on the facing page was taken while the spacecraft was above the Indian Ocean. It shows the crescent Earth from the South Pole at the left, through the "Roaring Forties" airmass of the South Atlantic Ocean to the west coast of Africa.

"These photographs," Phillips added, "contain significant technical data that are being evaluated to determine the techniques that will be employed for landmark identification for updating the Apollo navigational systems."

They also suggest what remarkable pictures cameras in space will produce during the next 10 years.

APPENDIX

Unlike aircraft, whose clean forms are shaped by the laws of aerodynamics, spacecraft have assumed a host of strange configurations. The laws that shape them are different: how to fold up to fit within a shroud during the quick, fierce passage through the atmosphere and then unfold in airless space; how to gather electrical power from the Sun; how to deploy antennas to beam data and photographs back to Earth; how to stay in a narrow band within temperature extremes; how to maintain pointing orientation in a weird environment that has no up or down; how to land in a rock-strewn crater. Only manned spacecraft, which must reenter the atmosphere, have of necessity avoided the spraddled, insectlike look of these machines that are truly something new under the Sun.

Spacecraft have paced the growth of much engineering technology in the last decade. Their special requirements for lightweight electronics, miniaturized computers, unique power supplies, and automated cameras and scientific instruments have forced engineers to break new ground. Spacecraft have functioned unattended in space for years, faithfully working and reporting back to Earth long after their designed mission has been accomplished. The following pages show many of the classes of spacecraft that took the photographs in this book.

The Television Infrared Observational Satellite (better known as Tiros) that pioneered in worldwide meteorological photography evolved into the ESSA satellite that NASA now delivers in orbit for the Environmental Science Services Administration. Nimbus is a larger, more sophisticated satellite designed to help develop instrumentation for long-range weather forecasting. The Applications Technology Satellites (ATS) carry meteorological and communications experiments to a 22 300-mile, Earth-synchronous orbit, from which they have provided spectacular full-Earth photographs.

Lunar and planetary spacecraft, a group distinct from Earth satellites, began with the Rangers, which were designed to radio back pictures as they approached the Moon to crash landings. The last three Rangers to be launched were completely successful, returning many thousands of excellent photographs.

The Mariner class of planetary explorers derived from Ranger technology, though needing improved guidance and communications systems. Mariner IV, man's first successful scout to Mars, set an excellent record of technical achievement: it completed a 7½ month flight to its planetary rendezvous, radioing its pictures and scientific data back 150 million miles. Then it continued on in lonely solar orbit for an additional 29 months. When it finally fell silent, more than 3 years after launch, it was not because of any failure but simply because it had consumed all of the onboard gas used to maintain its attitude in space.

Five Lunar Orbiters were flown, returning some of the most dramatic photographs in this volume. These craft were unique in that, instead of the vidicon television systems used in other unmanned spacecraft, they carried film cameras of very high resolution, along with automatic laboratories that developed and fixed each picture, and then electronically scanned it for transmission back to Earth.

Five Surveyors soft-landed on the Moon. On Earth command they sent back scores of thousands of photographs ranging from panoramic views to high-resolution closeups of nearby rocks and soil. Surveyor VII brought to an end America's highly successful automated exploration of the Moon in advance of man's first visit there.

The Mercury spacecraft that carried one man into near-Earth space were succeeded by the Gemini that carried two, and will be shortly followed by the Apollo spacecraft in which three astronauts will embark for the Moon. The shape, if not the size, of the first two classes of manned craft was similar, being designed for the 25 000-feet-per-second reentry velocity from Earth orbit. The Apollo spacecraft is slightly different, to suit the 36 000-feet-per-second velocity of reentry from a lunar trajectory.

201

Nimbus

The Nimbus satellite was conceived in 1958 as a relatively large satellite that would serve as a platform for research on advanced meteorological sensors. It operates in a Sun-synchronous polar orbit with its axis pointing toward Earth and its solar panels toward the Sun. This requires three-axis stabilization and rotating solar panels. Both Nimbus satellites launched to date were successful. Some characteristics of Nimbus I:

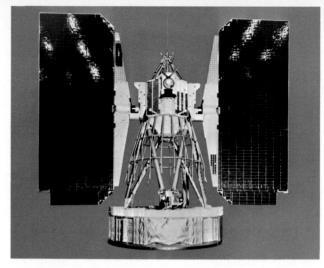

Weight at launch	912 lb
Power:	
Solar panels	165 watts
Batteries	224 ampere-hours
Communications:	
Tracking	035 watt
Wideband links	5 watts
Narrowband links	5 watts
Command receivers	2
Stabilization	3-axis torquing and damping system with horizon scanners, rate gyro, and inertia wheels
Pointing accuracy	±1°

Advanced vidicon systems	3 TV cameras and a tape recorder
Focal length	17 mm
Aperture	$f/4$ to $f/16$
Resolution	0.5 mile at 750 miles
Automatic picture transmission system	TV camera
Focal length	5.7 mm
Aperture	$f/1.8$
Resolution	1.4 miles at 750 miles
High-resolution infrared radiometer	Optical system, photo-detector, and tape recorder
Resolution	10 miles at 750 miles
Launch vehicle	Thor-Agena **B**

Tiros

Initiated by the Department of Defense in 1957, the Tiros Program was transferred to NASA shortly after its establishment. These satellites were designed to explore the feasibility of global weather observation from orbit. The first Tiros flew successfully in 1960 and the series went on to complete a string of 10 consecutive successes. Some specifications for Tiros VI were:

Weight at launch	280 lb
Power:	
Solar panels	33 watts
Batteries	12 ampere-hours
Communications:	
Dual-beacon transmitters	0.25 watt each
Dual TV transmitters	2 watts each
Crossed-dipole antenna	6-db gain
Stabilization	Spin stabilized with magnetic attitude control
Cameras (2):	
Focal length	5 mm
Aperture	$f/1.5$
Resolution	1.13 miles at 400 miles
Launch vehicle	Thor-Delta

ESSA

The ESSA meteorological satellite uses much of the basic Tiros technology, but has a number of important improvements. It flies in a Sun-synchronous polar orbit so as to maintain unchanging illumination of the Earth beneath it from day to day. The early Tiros satellites had their cameras pointing along the spin axis, which limited the coverage; the ESSA series has its cameras pointing radially so that each rotation of the spinning satellite brings the Earth into camera view. The satellites are precessed in orbit by electromagnetic coils so that the satellite appears as a wheel rolling along its orbital path. By these techniques, daily global coverage under proper lighting conditions is achieved. The ESSA series has a perfect record of six successful flights to date. Specifications for ESSA III:

Weight at launch	326 lb
Power:	
Solar panel	53 watts
Batteries	8 ampere-hours
Communications:	
Dual-beacon transmitters	0.25 watts each
Dual TV transmitters	5 watts each
Crossed-dipole antenna	6-db gain
Stabilization	Spin stabilized in cartwheel configuration, with magnetic attitude and spin control
Cameras (2):	
Focal length	5.7 mm
Aperture	$f/1.8$
Resolution	2 miles at 750 miles
Launch vehicle	Thrust-augmented Delta

ATS

The Applications Technology Satellite (ATS) is a multipurpose spacecraft designed to conduct research in synchronous Earth orbit at 22 300 miles above the Equator. In such an orbit, the satellite remains fixed relative to the subsatellite point on Earth. In addition to communications experiments, this satellite provided the first continuous meteorological observations of the world beneath it. Continuous observation will help detect violent but short-lived storms such as thunderstorms and tornadoes, whose life cycles may be less than the time for a low-orbit satellite to complete a single orbit. Of three ATS spacecraft flown, two were successful and one suffered a launch failure. Characteristics of ATS–III:

Weight:	
At launch	1574 lb
After kick-motor firing	798 lb
Power:	
Solar panel	189 watts
Batteries	6 ampere-hours
Communications:	
Dual transmitters	4 watts each
Dual transmitters	12 watts each
Dual transponders	
Stabilization	Spin stabilized with 4 monopropellant thrusters, each of 5-lb thrust
Multicolor spin-scan camera:	
Focal length	375 mm
Aperture	$f/3$
Resolution	1.9 miles at 22 300 miles
Image-dissector camera:	
Focal length	49.2 mm
Aperture	$f/2$
Resolution	6 miles at 22 300
Launch vehicle	Atlas-Agena

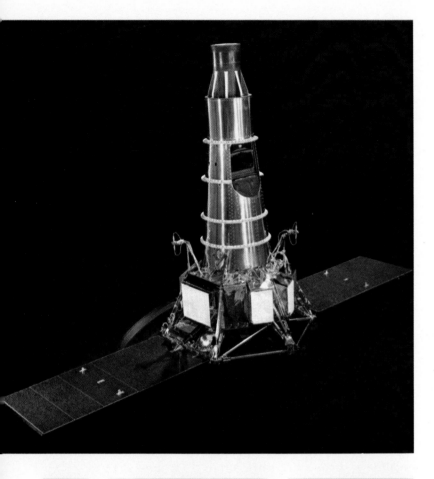

Ranger

Ranger was the first U.S. spacecraft designed to make sophisticated observations of another celestial body—the Moon. Nine Rangers were built in all. Of these, the first two were test vehicles for Earth orbit and experienced launch failures. Seven were launched to the Moon. The first three were designed to take moderate-resolution photographs during approach and to rough-land a seismometer capsule; they failed due to a combination of launch-vehicle and spacecraft problems. The last four Rangers were exclusively devoted to detailed lunar photography with a battery of six vidicon cameras. An arcing problem during launch rendered Ranger VI's cameras inoperable. However, Ranger VII, VIII, and IX completed perfect missions and returned thousands of excellent photographs of the lunar terrain. Specifications for the final (Block III) Ranger:

Weight at launch .... 802 lb
Power:
    Solar panels .... 200 watts
    Spacecraft batteries .... 90 ampere-hours
    TV batteries .... 40 ampere-hours
Communications:
    Dual spacecraft transmitters ... 3 watts each
    Dual TV transmitters .... 60 watts each
Midcourse propulsion .... Liquid monopropellant of 50-lb thrust

Guidance and control:
    Inertial reference .... 3-axis gyros
    Celestial reference .... Earth and Sun sensors
    Attitude control .... Cold-gas jets
Cameras (6):
    Focal lengths .... 76 and 25 mm
    Apertures .... $f/1$ and $f/2$
    Exposures .... 2 and 5 msec
Launch vehicle .... Atlas-Agena

Mariner

Mariner spacecraft were designed to reconnoiter the planets Mars and Venus. Mariner I experienced a launch-vehicle failure, but Mariner II made the world's first successful flight to another planet, passing Venus 109 days after launch at a distance of 21648 miles, and recording numerous scientific observations. Mariner III also experienced a launch failure, but Mariner IV made the world's first successful flyby of Mars at a distance of 6118 miles. The most recent of the Mariner series, Mariner V, successfully observed Venus from a distance of 2131 miles in October 1967, but, like its Mariner II predecessor, carried no camera. Specifications for Mariner IV:

Weight at launch	574 lb
Power:	
Solar panel near Earth	700 watts
Solar panel near Mars	325 watts
Batteries	1200 watt-hours
Communications:	
Transmitter	10 watts
Antenna	23-db gain
Bit rate near Earth	33.3 bps
Bit rate near Mars	8.33 bps
Guidance and control:	
Inertial reference	3-axis gyros
Celestial reference	Sun and Canopus sensors
Attitude control	Cold-gas jets
Propulsion: Liquid monopropellant.	50-lb thrust
TV camera:	
Optical system	Cassegrain telescope
Focal length	30.48 cm
Aperture	$f/8$
Field	1.05° by 1.05°
Launch vehicle	Atlas-Agena

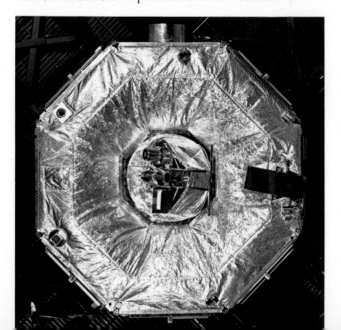

Lunar Orbiter

Lunar Orbiter was designed to operate in conjunction with Surveyor. By photographing Surveyor sites from orbit, the detailed Surveyor findings could be extrapolated to other areas that appear identical in orbital photography. All five Lunar Orbiters were successful. They photographed all potential Apollo landing sites to resolutions down to 1 meter. Having completed the Apollo photographic requirement, the last two Orbiters photographed the entire front face of the Moon at resolutions of some 10 times better than possible from Earth, completed photography of the far side at resolutions somewhat better than obtainable of the front side from Earth telescopes, and obtained detailed photography of several dozen lunar features of particular scientific interest. Some Orbiter specifications:

Weight at launch	856.71 lb
Power:	
Solar panels	450 watts
Batteries	12 ampere-hours
Communications:	
Telemetry transmitter	0.48 watt
TV transmitter	10 watts
Guidance and control:	
Inertial reference	3-axis gyros
Celestial reference	Sun and Canopus sensors
Attitude control	Cold-gas jets
Propulsion:	
Liquid bipropellant	100-lb thrust
Cameras (2):	
Focal lengths	60.96 cm and 80 mm
Aperture	$f/5.6$ (both)
Resolution	1 and 8 m
Launch vehicle	Atlas-Agena

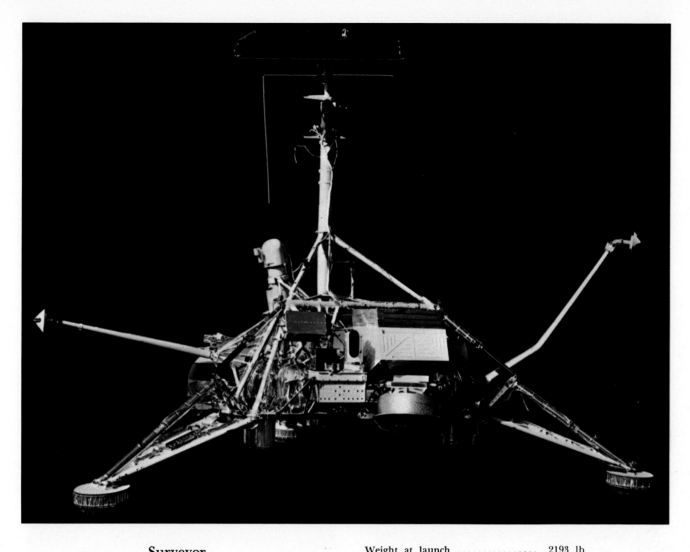

Surveyor

The Surveyor spacecraft series was conceived in 1959 to soft-land scientific experiments on the Moon and carry out initial surface investigations. It was an ambitious undertaking at the time, for it relied on developing the new technology of closed-loop, radar-controlled landings, and, in addition, required the successful development of the world's first hydrogen-oxygen rocket stage, the Centaur. Of seven spacecraft, five were completely successful; two experienced in-flight failures prior to landing. The Surveyor not only studied four potential Apollo landing areas, but on its last mission visited the scientifically interesting crater Tycho. In addition to relaying thousands of surface photographs to Earth, Surveyor measured surface properties by manipulating the lunar soil with a mechanical arm and conducted chemical analysis of the lunar material. Typical Surveyor specifications:

Weight at launch	2193 lb
Landed weight	625 lb
Power:	
Solar panel	90 watts
Batteries	230 ampere-hours
Communications:	
Dual transmitters	10 watts each
Planar-array antenna	27-db gain
Guidance and control:	
Inertial reference	3-axis gyros
Celestial reference	Sun and Canopus sensors
Attitude control	Cold-gas jets
Terminal landing	Automated closed loop, with radar altimeter and Doppler velocity sensor
Propulsion:	
Main retrorocket	Solid fuel of approximately 9000-lb thrust
Liquid bipropellant vernier retrorockets	Throttleable between 30- and 102-lb thrust each
TV camera:	
Focal length	25 or 100 mm
Aperture	$f/4$ to $f/22$
Resolution	1 mm at 4 m
Launch vehicle	Atlas-Centaur

Mercury

The Mercury spacecraft was designed to conduct the first U.S. experiments with man in Earth orbit. This spacecraft was constrained in weight by the capability of a modified Atlas launch vehicle and pioneered in lightweight manned spacecraft capable of reentry through the atmosphere. During the four successful orbital Mercury missions, astronauts Glenn, Carpenter, Schirra, and Cooper accumulated 53 hours of space-flight experience. Typical Mercury specifications:

Weight at launch	3000 lb
Power: Batteries	590 ampere-hours
Stabilization	Redundant semiautomatic and manual systems using hydrogen peroxide thrusters
Crew	1 astronaut
Cameras flown	70-mm modified Hasselblad; 35-mm zodiacal-light camera; 16-mm Maurer cine camera
Launch vehicle	Atlas

Gemini

This program was planned to investigate longer duration manned flight and the practicality of rendezvous and docking in space. In 10 missions, each with a 2-man crew, 1940 man-hours of flight time were accumulated. Included were a flight lasting 2 weeks, six rendezvous missions, four docking missions, use of a docked propulsion stage to maneuver to high altitudes, 5 hours of extravehicular activity, and the conduct of numerous scientific experiments. Some characteristics of the Gemini spacecraft:

Weight at launch	7100 lb
Power:	
Batteries	160 ampere-hours
Fuel cells	6570 ampere-hours
Stabilization	Redundant semiautomatic and manual systems using bipropellant thrusters
Crew	2 astronauts
Cameras flown	70-mm modified Hasselblad; 70-mm Maurer; 16-mm Maurer cine cameras; and special scientific cameras
Launch vehicle	Titan II

208

Apollo

The Apollo command and service module (CSM) shown here is designed to carry three astronauts into lunar orbit from which two of them will descend to a landing in another spacecraft, the lunar module. After the lunar landing and subsequent rendezvous in lunar orbit, the CSM will inject itself into an Earth-return trajectory. On reaching Earth, the command module alone will reenter the atmosphere at a velocity of about 36 000 feet per second, some 11 000 fps faster than reentry from Earth orbit. The command module contains advanced guidance and control equipment for accurate maneuvering into lunar orbit and into the narrow-entry corridor upon return.

Weight at launch	64 500 lb
Command module	13 000 lb
Service module	11 800 lb
Service module propellant	39 700 lb
Power:	
Batteries	1250 ampere-hours
Fuel cells	19 300 ampere-hours
Stabilization	Dual systems, one on each module, semiautomatic and manual, using liquid bipropellant thrusters
Crew	3 astronauts
Cameras flown	35-mm Maurer; 16-mm Millikan
Launch vehicle:	
Earth-orbital missions	Uprated Saturn I
Lunar missions	Saturn V

INDEX

A

Adams, John B., 74
Advanced Vidicon Camera System, 19, 22, 30
Aerobee sounding rocket, 4
Africa, photographs, 156, 157, 169, 170, 171, 176, 177, 189, 191, 192
Agena vehicle, 137, 172, 173, 174, 182, 183, 184, 185, 195, 205
Agriculture, U.S. Department of, 152
Air Force Missile Test Center, 4
Airglow layer, 140
Aldrin, Edwin E., Jr., 150, 183, 184, 185
Alexiou, Arthur, 155
Allenby, Richard J., 157
Alley, C. O., 80
Ames Research Center, 86, 106
Andes Mountains, cloud cover, 167, 179
Anscochrome film, 152
Antarctica, 22
Apollo landing zone studies, 74, 92–93, 206, 207
Apollo Lunar Exploration Office, 124
Apollo Program, 53, 73, 78, 80, 93, 137, 154, 167, 173, 182, 196, 197, 198, 201, 209. *See also* Apollo landing zone
Applications Technology Satellite Program, 2, 3, 33–34, 201, 203. *See also* ATS
Arabian Peninsula, photographs, 146, 190, 192
Argonne National Laboratory, 72
Aristarchus, lunar crater, 122, 123, 124, 125
Armstrong, Neil A., 172
Atlas flight, 4
Atlas Mountains, photographs, 139, 208
Atlas-Agena, 205, 206
Atlas-Centaur, 207
Atomic Energy Commission, 72
ATS:
    I, 33, 34, 36, 37
    III, ii, 3, 203
Aurora 7, 139
Australia, photograph, 195
Automatic Picture Transmission System, 3, 14, 26

B

Bahamas, photographs, 143, 155
Baja California, photographs, 166, 178
Baldwin, Ralph B., 56
Batterson, Sidney A., 65
Bellcomm, Inc., 52, 81, 124
Belyayev, Pavel, 150
Berg, Otto, 4
Berry, Dr. Charles A., 185, 191
Bessemoulin, Jean, 18
Bogart, Frank A., 184
Borman, Frank, 144, 164
Boyer, William J., 102
Brinkmann, John R., 158
Bristor, Charles L., 32
Brunk, William E., 88
Bryson, Robert P., 119
Bureau of Commercial Fisheries, 143
Burke, Joseph R., 2, 33
Butler, Herbert I., 30
Bykovsky, Valery, 137

C

California coast, photograph, 153
California Institute of Technology, 82, 131, 132, 133
Canada, Department of Transport, 10
Canary Islands, cloud cover, 168
Canopus, x, 181
Carpenter, Scott, 137, 138, 208
Cartography, 22
Centaur vehicle, 207
Cernan, Eugene, 173, 174, 178, 180
Chaffee, Roger, 151
Chandler, C. E., 82
Chidester, A. H., 147
Choate, Raoul, 77
Christaller's Central Place Theory, support, 159
Christensen, Elmer M., 67
Christensen, Everett E., 189
Churning air, characteristics, 28
Clark, John F., 37
Clary, Maurice C., 66, 67
Cloud, James D., 71

Clouds:

    pattern interpretation, 21
    pictures from ESSA satellites, 23–32
    pictures from Nimbus, 14–18
    pictures from Tiros, 6, 12–13
    See also Vortex structures
Cobra Head, lunar feature, 124
"Cold line" meteorological link, 10
Collins, Michael, 174–175
Color analysis, lunar soil, 127
Colorado River, photograph, 145
Colwell, Robert N., 152
Commerce, U.S. Department of, 152, 156
Conrad, Charles, Jr., 137, 152, 181, 182, 185, 191, 195
Cooper, Gordon, 137, 140, 141, 152, 191, 208
Copernicus, lunar crater, 88, 89, 102, 116–117
Cordillera Mountains, lunar feature, 110, 112
Cortright, Edgar M., v, x
Cosmos CXLIV, 11
Crabill, Norman L., 101
Craters, lunar, *see* by name
Cunningham, Newton W., 94, 95
Currie, D. G., 80
Cyclonic storm life cycle, 36
Cyrillus, lunar crater, 102

D

Daily global weather coverage, 30
Damoiseau, lunar crater, 101
Dankanyin, R. J., 70
Darcey, Robert J., 34
Davies, D. A., 26
Day, Leroy E., 167
Debus, Kurt H., 154
Defense, U.S. Department of, 202
Delta vehicle, 202, 203
de Wys, Jane Negus, 72
Donlan, Charles J., 99
Douglas Advanced Research Laboratory, 50
Draley, E. C., 102
Dwornik, Stephen E., 51

☆ U. S. GOVERNMENT PRINTING OFFICE: 1968 O–292–583